Cathodic Protection for Reinforced Concrete Structures

Cathodic Protection for Reinforced Concrete Structures

Paul M. Chess

CRC Press
Taylor & Francis Group
Boca Raton London New York

CRC Press is an imprint of the
Taylor & Francis Group, an **informa** business

CRC Press
Taylor & Francis Group
6000 Broken Sound Parkway NW, Suite 300
Boca Raton, FL 33487-2742

First issued in paperback 2021

ISBN 13: 978-1-03-209451-9 (pbk)
ISBN 13: 978-1-138-47727-8 (hbk)

This book contains information obtained from authentic and highly regarded sources. Reasonable efforts have been made to publish reliable data and information, but the author and publisher cannot assume responsibility for the validity of all materials or the consequences of their use. The authors and publishers have attempted to trace the copyright holders of all material reproduced in this publication and apologize to copyright holders if permission to publish in this form has not been obtained. If any copyright material has not been acknowledged please write and let us know so we may rectify in any future reprint.

Publisher's Note
The publisher has gone to great lengths to ensure the quality of this reprint but points out that some imperfections in the original copies may be apparent.

Library of Congress Cataloging-in-Publication Data

Names: Chess, Paul M., author.
Title: Cathodic protection for reinforced concrete structures / Paul M. Chess.
Description: Boca Raton : Taylor & Francis, a CRC title, part of the
Taylor & Francis imprint, a member of the Taylor & Francis Group, the
academic division of T&F Informa, plc, [2019] | Includes bibliographical
references and index.
Identifiers: LCCN 2018038767 (print) | LCCN 2018036500 (ebook) |
ISBN 9781138477278 (hardback : acid-free paper) | ISBN
9781351045834 (ebook) | ISBN 9781351045827 (Adobe PDF) | ISBN
9781351045810 (ePub) | ISBN 9781351045803 (Mobipocket)
Subjects: LCSH: Buildings, Reinforced concrete—Protection. |
Reinforced concrete—Cathodic protection.
Classification: LCC TH1501 .C478 2019 (ebook) | LCC TH1501
(print) | DDC 624.1/83410289—dc23
LC record available at https://lccn.loc.gov/2018038767

Visit the Taylor & Francis Web site at
http://www.taylorandfrancis.com

and the CRC Press Web site at
http://www.crcpress.com

Contents

Author

Paul M. Chess was formerly the Managing Director of the largest specialist manufacturer of products for cathodic protection of concrete in the world, Cathodic Protection International. He is currently the Managing Director of Corrosion Remediation Limited.

CHAPTER 1

The Corrosion Process in Reinforced Concrete

The State of the Art

INTRODUCTION

The use of steel in concrete has increased massively in the last century, and it is now one of the most commonly used building materials on earth (second to water in usage by mankind). The biggest durability problem with structures of this material is corrosion of the steel reinforcement. The cause of this corrosion that is most difficult to remedy is the presence of chloride in the concrete, which allows the steel to corrode at a rapid rate. Because of the huge economic significance of this problem over the last 50 years, there has been almost continuous research in many countries to categorise the problem.

There have been significant attempts to improve the actual life of a reinforced concrete structure by changing the material properties of both the reinforcement and concrete. Secondly, there have been changes to cover depths of the concrete. Relatively recently, certain countries, such as Saudi Arabia, have been installing cathodic prevention on a large scale to prevent corrosion from initiating by forcing the steel reinforcement to remain as a cathode.

As of yet, there has not been a thorough look at changing the structural design of the steel reinforcement in order to minimise the risk of premature corrosion, although this practice has been ongoing for many years in the case of steel structures. This design change would be to minimise the formation of anodic sites where preferential corrosion occurs.

The aim of this chapter is not to repeat the accepted wisdom of previous textbooks but to look at the results of recent research and contemplate an alternative view of the corrosion process.

THE CORROSION PROCESS

How steel corrodes in concrete is of significant importance when trying to understand the likely mechanisms and possible rates. If a clean-surfaced, grit-blasted, rebar is put in a saline, pH-neutral solution, then there will be a rapid browning over the whole surface area. A bright yellowish brown oxide will form within minutes and gradually become darker as the exposure time increases. This is commonly referred to as microcell corrosion, as it is appearing evenly over the whole surface of the rebar. As the exposure time increases, the rate of corrosion reduces, as the oxide layer provides a barrier to ionic, atomic and gaseous transport.

In reinforced concrete structures, when the contaminated concrete cover is removed after many years, the corrosion process is not so simple. There can be areas where corrosion appears to be general with uniform section loss (refer to Figure 1.1). This tends to be in areas with lower cover depth and relatively dry carbonated concrete.

There can also be corrosion pitting where there is a significant loss of section at defined locations (refer to Figure 1.2 and, for the

FIGURE 1.1 Example of significant general (uniform) corrosion on a marine structure.

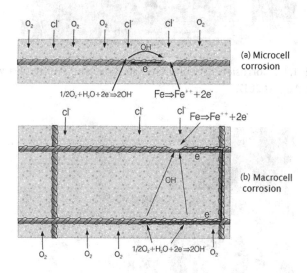

FIGURE 1.2 Schematic of micro- and macro-corrosion mechanism.

result of this loss of section, to Figure 1.3). This is most commonly found in wet and high-cover parts of a structure. When exposed, the oxides found in these pits often have some black and green colourations and can smell of chlorine gas or hypochlorous acid, both recognisable from swimming pools. In this case, the oxidation product does not appear to provide a protective barrier and indeed may be providing a poultice where low pH and aggressive ions can remain (Figure 1.4).

FIGURE 1.3   Corrosion of reinforcing steel causing failure of a waffle deck in August 2017.

FIGURE 1.4   Corroded steel reinforcement showing pitting corroded steel.

So it appears that there are at least two distinct corrosion mechanisms as shown schematically in Figure 1.2. Several researchers (Green, 2017) have looked at different hypotheses for pit formation, and presently, the most favoured is the transitory complex model where the chloride ions form a soluble compound, which moves away from the anodic site at the base of the pit. Away from the corrosion site, the iron hydroxide precipitates re-releasing the chloride ions. This mechanism has been observed (Angst et al., 2011) on actual reinforced concrete specimens with initial pit formation and the corrosion product deposited on the rebar a small distance away. It seems likely that there are several stages to pit formation and the actual mechanisms will be varying through the corrosion process. What the above findings demonstrate is that the corrosion process is much more complex than that in a solution and thus cannot be modelled using the standard electrolytic formulas such as the Stern Geary equation. This is an equation that relates quantitatively the slope of a polarisation curve in the vicinity of the natural corrosion potential to the corrosion current density and is discussed in more detail in Chapter 7.

Another variable that has only recently started to be explored is the differences in the composition and structure of the reinforcement itself having a significant effect on corrosion initiation and propagation. For example, it has been found that manganese sulphide inclusions can preferentially be corroded in chloride-rich reinforced concrete, and this could initiate or fully form pits. Other observations which have been made are that the form of hot roll processing favouring, respectively, pearlite and lower- or upper bainite also has a significant influence on the form and extent of the corrosion process. Recently, the entire steel–concrete interface has been looked at more closely and a review published (Geiker, 2017).

The initial iron oxidation reaction is anodic and involves the loss of electrons, which can be represented by

$$Fe = Fe^{2+} + 2e^-.$$

This initial reaction is then followed by several further reactions leading to a more ionically charged atom and further electron production. The important point is that there are electrons being produced and that these are travelling through the steel matrix to the cathode on the steel surface, which is a non-corroding area where they combine with oxygen and water in the cathodic reduction reaction. In the microcell-type corrosion, this is likely to be with the anode and cathode as part of the same grain, which is typically from 0.25 mm down to 0.025 mm. This is because steel is an alloy with components that have different potentials.

THE AMOUNT OF CHLORIDE REQUIRED TO INITIATE CORROSION

This has been exhaustively evaluated over many years, and a review of the many experimental procedures undertaken was published (Angst, 2009). In this review, the parameters that affect the onset and amount of corrosion are discussed and found to include:

- Steel–concrete interface
- Concentration of hydroxide ions in the pore solution (pH)
- Electrode potential of the steel
- Binder type
- Surface composition of the steel
- Moisture content of the concrete
- Oxygen availability at the steel surface
- Water-to-binder ratio
- Electrical resistivity of the concrete
- Degree of hydration
- Chemical composition of the steel

- Temperature
- Chloride source
- Type of cation accompanying the chloride ion
- Presence of other, inhibiting or otherwise, species

With all these variables, it is perhaps not surprising that there would be a significant range in the critical chloride level. However, what was not anticipated was the huge range in chloride levels where corrosion has been initiated in different research studies, and these ranged from 0.04% to 8.4% chloride by weight of cement, which is a 21,000% difference (Angst, 2009). This massive difference is to a certain extent replicated in real structures where the range was from 0.1% to 1.95% chloride by weight of cement, which is a 1,950% difference. These huge disparities mean that deterministic life expectancy modelling, which is applied to many important and vulnerable structures and has been refined and honed over many years, is inherently flawed. This is because all these models take an arbitrary chloride level and assume that the chloride will move through the concrete at a certain diffusion rate following Fick's law until it builds to a sufficient concentration level when it depassivates the steel reinforcement. If the level of chloride required to initiate corrosion is extremely variable, then this model cannot be followed with confidence. A more effective method may be to move to a probability distribution for the threshold. Later in this chapter, the assumption that a diffusion model is the way that charged ions move around a structure is also looked at more closely.

THE EFFECT OF THE STRUCTURE'S SHAPE ON CORROSION

Recently, an experiment was undertaken (Angst and Elsener 2017) where the specimen size was varied in the same experimental procedure. It was found with exactly the same experimental conditions and procedures that reinforced concrete samples with 1 cm, 10 cm and 100 cm lengths required very different chloride concentrations to initiate corrosion. No corrosion was found at more than 2.4% chloride by weight of cement for the 1 cm sample, with the 10 cm sample initiating corrosion at an average of 1.5% chloride by weight of cement, and the 100 cm sample initiating corrosion at an average of 0.9% chloride by weight of cement.

This has profound implications in that it means that the steel reinforcement layout has a large effect on the results obtained both in experimental procedures and also more importantly in real structures. Even more worryingly, these large real structures are likely to perform substantially worse than may be predicted by laboratory testing with small samples. The reason for this behaviour was conjectured by Angst and Elsener (2017) to be that the bigger the area of the steel reinforcement, the more likely there was to be inhomogeneities on the surface of the steel where this corrosion process can prosper. Whatever the reason for the results, this experiment shows that the presence and layout of steel in the concrete has a dramatic effect on the corrosion process, and this should be considered important. Most of the durability experiments that have been undertaken to date have neglected this fact. With this experiment and some further evidence which is outlined below, it should be understood that concrete on its own and steel reinforced concrete do not necessarily behave in the same way.

IONIC MOVEMENT IN CONCRETE

Modern structures in aggressive environments are increasingly being engineered to achieve a life expectancy based on the impenetrability of the concrete to chloride ions. These predictions are normally based on Fick's laws of diffusion even though it is commonly known that this is an over simplification of the transport situation in concrete. It is widely known that advection (capillary suction) and migration can occur in concrete, and its common use is probably linked to its simplicity and ability with certain situations to adequately predict reality with adequate precision.

Fick's law is a concentration diffusion mechanism and assumes that transport is occurring in concrete in exactly the same way as in an aqueous bath. This is probably a reasonable assumption for a simple concrete sample, but it is not so reasonable when there are electrical potential differences in the structure, such as between two steel bars in a reinforced concrete sample and the huge number of bars at different depths and orientations in a real structure. The effects of this difference in potential will be to create an electrical field that promotes the movement of ions by electro-osmotic flow from the anode to the cathode or vice versa depending on their charge and solubility.

It has been known for more than 200 years (Reuss, 1809) that water could be made to percolate through porous clay diaphragms through the application of an electric field. The clay in this case behaves in the same way as concrete in that the particles acquire a surface charge when in contact with an electrolyte. Water is not electrically neutral and is a polar molecule. This means there is a small net negative charge near the oxygen atom and partial positive charge near the hydrogen atoms. So when there is a net charge in the electrolyte, the water molecules align themselves in a polarised way on the solid particles, which, in this case, will be the pore walls of the concrete. The immobile surface charge in turn attracts a cloud of free ions of the opposite sign creating a thin layer of mobile charges next to it. This electric double layer is typically 1–10 nm wide and is commonly called the Debye layer (refer to Figure 1.5). In the presence of an external electrical field, the Debye layer acquires a momentum, which is then transmitted to adjacent layers of the pore solution through the effects of viscosity. This is the process of electro-osmosis.

Evidence for this mechanism in reinforced concrete is provided by the fact that it has been known for many years that the application of an external direct current can dramatically alter the rate at which ions, such as chlorides, can move through concrete, and this technique is used to make rapid assessments of the likely permeability of concrete samples. In a test such as the commonly

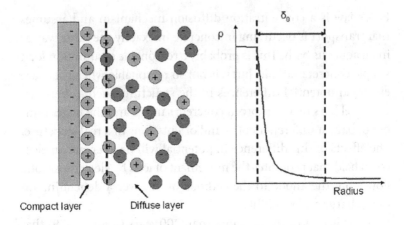

Compact layer   Diffuse layer

FIGURE 1.5   Electrical double layer.

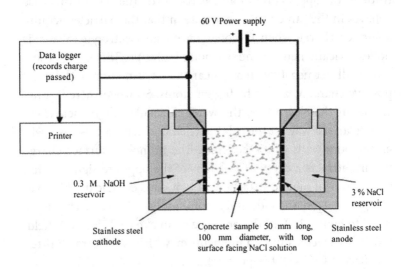

FIGURE 1.6   Drawing of Coulomb test layout.

called Coulomb test (ASTM C1202, 1997), the thickness of the sample is 50 mm and the test is run for 6 h at 60 V. The amount of chloride moved is equivalent to that which would be expected in 50 years in normal conditions, so that constitutes a rate increase of 73,000 times relative to straight diffusion as shown in Figure 1.6.

In the last few decades, electrochemical treatment of reinforced concrete such as electrochemical chloride removal and impressed current cathodic protection of reinforced concrete structures have become increasingly commonplace and are now being routinely installed in many projects around the world. In both of these processes, there are significant flows of ions at far higher rates than would be predicted by diffusion calculations.

So the question arises, is this mechanism of electro-osmosis relevant at the potential differences that exist in actual structures, and if so, does this then have a significant effect on the durability? For these questions to be answered, some testing on actual structures is reported below:

STUDIES OF CORROSION ON STRUCTURES

A seasoned investigator (Gjorv, 2014) looked at many marine structures in Norway, including harbour wharves. He found that different designs of reinforced concrete wharf in exposed marine conditions had massively different durability outcomes despite these being built at the same time with the same concrete, steel reinforcement and cover depths. He concluded that the reinforced concrete structures exposed to a chloride-containing environment would develop a complex system of galvanic cell activities along the embedded steel. In such a system, the more exposed parts of the structure such as the deck beams would always absorb more chlorides and hence have anodic areas, while the less exposed areas, such as the slab sections in between, would act as cathodes. He found that wharves constructed without these beams were massively more durable with no corrosion on the bottom surface after 70 years.

Gjorv looked at a recent cruise terminal marine wharf as shown in Figure 1.7 and found that the steel in the reinforced deck beams was corroding within 8 years after construction despite the concrete having a water/binder ratio of 0.45, a 28-day compressive strength of 45 MPa, silica fume addition and an average cover depth of 50 mm. No sign of this corrosion was given by half-cell potential mapping, cracking or staining of the concrete. From

FIGURE 1.7   Cruise terminal in Norway.

conventional wisdom on the performance of steel in reinforced concrete, there should be no corrosion after such a short time.

Another seasoned investigator (Melchers, 2017) has also looked at many reinforced concrete structures and came to the following conclusions:

1. Very high levels of chloride concentrations on reinforcing bars does not necessarily 'cause' other than very minor corrosion, and often with no visual evidence of corrosion at all, for many years, often decades.

2. For high-quality concretes, the onset of serious, damaging corrosion is considerably later in time than any measure of corrosion initiation.

3. The causative mechanism(s) for the commencement of serious reinforcement corrosion remain unclear but do not appear to be the same as those for corrosion initiation.

4. Very severe localised reinforcement corrosion without obvious external signs of corrosion such as rust staining and concrete cracking or spalling can and does occur, but clear explanations for such corrosion remain outstanding.

What these studies demonstrate is that there is a great unpredictability in the way that structures behave in a similar environment. This leads to the only possible conclusion in that it is the design of the reinforced concrete structure, which has the most significant effect on its corrosion durability in an aggressive environment. So if a structure is designed so that there is a similar potential between all the steel in the reinforced concrete elements, then the rate of corrosion will be very low even at high chloride levels. This conclusion implies that there is a significant and probably dominant electrochemical effect on the corrosion process, and the preoccupation with chloride levels is incorrect.

INCIPIENT ANODES

It has been commonly found in repairing chloride-damaged structures by patch repair with fresh mortar or concrete that the steel in nearby contaminated concrete then suffers from accelerated corrosion (Broomfield, 2007). The reason for this corrosion occurring in these particular types of patch repairs is that the previously corroding anodic areas become cathodic as they are covered with the high alkalinity mortar while the previously cathodic areas in the surrounding structure have then become anodic (refer to Figure 1.8).

A consideration of this is that the physical environment is not changed at the previous cathodic location where corrosion is now occurring. Although the immediate physical environment is unchanged, the only thing that has changed is the potential difference is altered. Thus, this is the deciding factor in whether corrosion occurs or not. With this potential difference, there could be electro-osmotic movement or electromigration, and this could after a very short time alter the composition of the concrete at the steel interface. The other factor that the incipient anode effect demonstrates is that a potential difference change of less than half a volt (the difference between a typical potential value of steel in clean concrete and a typical potential value of steel in

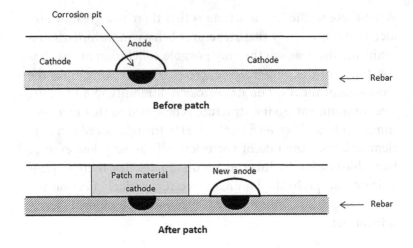

FIGURE 1.8 Formation of an incipient anode.

chloride-contaminated concrete) is enough to trigger this change from no corrosion to corrosion.

CONCLUSIONS

The traditional way of looking at the performance of reinforced concrete structures has been to look at the performance of the concrete and optimise this by both increasing the concrete's diffusion resistance and increasing the cover. What has not been fully recognised to date is that when steel reinforcement is added to the concrete, it can fundamentally change the behaviour of the concrete in that ions are moved not just by concentration diffusion but also by electro-migration. This additional movement mechanism has many fundamental implications for the study of durability and also the best design practice for structures. One of the most surprising things which has become evident is the speed of movement around a structure in the presence of an electric field, perhaps this is because when we look at concrete, we have the impression it is inert and unmovable.

The presence of electro-migration has not been considered by any of the international standards with which reinforced concrete structures are routinely designed to comply with. There could be two ways to improve the likely durability of the structure.

The first method to increase the durability as discussed by Gjorv (2014) is to make the steel reinforcement as iso-potential as possible. In practice, with a jetty, for example, this could mean not having beams supporting the deck but having a thicker deck slab which can withstand more loading, protecting the columns at transition sites between the water and the air and also between the deck and the exposed column.

Another way to increase durability could be to deliberately introduce areas that will be sacrificial to the main structure. These locations could be non-structural and replaceable. The design of such areas would have to be carefully researched and trialled before becoming an accepted means of corrosion control. Obviously, a cathodic protection system will also be effective in this role.

REFERENCES

Angst U, B Elsener, C Larsen, O Vennesland, Chloride induced reinforcement corrosion: rate limiting step of early pitting corrosion, *Electrochemica Acta*, 56, (2011), 5877–5889.

Angst U et al., Critical chloride content in reinforced concrete – a review, *Cement and Concrete Research*, 39, (2009), 1122–1138.

Angst U, B Elsener, The size effect in corrosion greatly influences the predicted lifespan of concrete infrastructures, *Science Advances*, 3, (2017), e1700751.

ASTM C1202–97, *Standard Test Method for Electrical Indication of Concrete's Ability to Resist Chloride Ion Penetration*, (1997), ASTM International, West Conshohocken, USA.

Broomfield J, *Corrosion of Steel in Concrete*, (2007), Taylor and Francis, Abingdon, UK, 2nd edn, p. 120, ISBN 0-415-33404-7.

Geiker A, M Gehlen et al., The steel concrete interface, *Materials and Structures*, 50, (2017), 143.

Gjorv O, *Durability Design of Concrete Structures in Severe Environments*, (2014), CRC Press, Boca Raton, USA, ISBN 13-978-1-4665-8729-8.

Green W, F Collins, M Forsyth, *Reinforced Concrete Corrosion Protection Repair and Durability*, (2017), Australasian Corrosion Association, Blackburn, Australia, ISBN 978-0-646-97456-9.

Melchers R E, *Reinforced Concrete Corrosion Protection Repair and Durability*, (2017), Australasian Corrosion Association, Blackburn, Australia, ISBN 978-0-646-97456-9.

Reuss F, *Mem Soc Imperiale Naturilists de Moscow*, (1809), Moscow University, *Russia*, 2 edn, 327 F.

CHAPTER 2

History of Cathodic Protection in Reinforced Concrete

It is commonly written that Sir Humphrey Davy was the first investigator of the practice of preferential corrosion of metals, but there is evidence that the Romans (Baeckmann, 2009) got there first and protected their copper fastenings with lead caps on their marine craft. Of course Davy was important, but it was Robert Kuhn who can be considered the father of impressed current cathodic protection (ICCP) as shown in Figure 2.1, while also developing, by trial and error, in 1928 a protection criterion of −850 mV with respect to a copper/copper sulphate electrode which is still in widespread use today. After the second world war, the use of cathodic protection (CP) became widespread in both the United States and Northern Europe on pipelines, tanks and other ferrous objects, first mainly with galvanic anodes, but as time progressed, more commonly with impressed current systems. The present situation is that all commercial shipping, gas pipelines and oil lines are universally protected, and there is CP

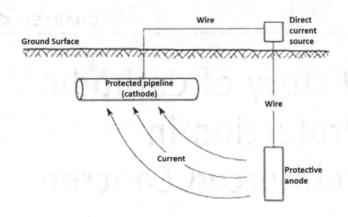

FIGURE 2.1   Schematic of an ICCP system.

routinely used on a whole host of other products such as water heaters, outboard engines and storage tanks.

Steel reinforced concrete has become one of the most commonly used construction materials in the world. In most applications, its unique blend of properties has provided a durable and cost-effective solution. However, in aggressive conditions, in some circumstances, it has proven durable and in others not (Melchers, 2017). Where the required design life has not been achieved, various repair strategies have been developed but most have been ineffective. In the United States, a particular problem was found on reinforced concrete bridge decks, which had been salted for ice control regularly over the winter months (refer to Figure 2.2). These decks had no waterproofing membrane and were roughly 20 years old. Sometimes, the running surface was concrete and other designs had asphalt as a top surface. In both these designs, the steel reinforcement in the concrete was actively corroding. This provoked the research described below.

ICCP was first used in laboratory trials on reinforced concrete in the United States (Stratfull, 1959). Despite cynicism that CP would not be suitable in such a high-resistivity environment, Stratfull (1974) persisted with full-scale installations to stop premature deterioration of deicing salt-affected bridge decks in

California as shown in Figure 2.2. The first systems borrowed heavily from pipeline CP technology in that they used ground beds such as silicon iron anodes in a graphite backfill which they placed on the concrete deck surface (refer to Figure 2.3).

FIGURE 2.2 Damage to an American bridge deck. (Courtesy of John Broomfield.)

FIGURE 2.3 Silicon iron anodes being covered by a carbonaceous backfill. (Courtesy of John Broomfield.)

The other commonly used system was a conductive polymer poured into slots cut in the upper deck surface with a metallic primary anode also in the slot (refer to Figure 2.4).

It quickly became apparent with this slot system that because of the closeness of the anode and cathode (the part passing the current and the steel reinforcement) and the high resistance of the electrolyte (the concrete), the maximum anode spacing to get a reasonable distribution of current was in the order of 1 foot (330 mm). It also became apparent that after the structural repair of the reinforced concrete, the anode was the most expensive part of the ICCP installation. Interestingly, no real anode favourite emerged in the early years, and it was split into these two types which was decided by the presence or absence of an asphalt cover. By the end of the 1970s, it was concluded by the American authorities that CP was a reliable repair technique even in the presence

FIGURE 2.4 Placing conductive polymer anode in slots with a carbon fibre primary anode. (Courtesy of John Broomfield.)

of high levels of chloride for the upper surface of bridge decks and could prevent the ongoing corrosion of the steel reinforcement. They did not consider that the technology was mature enough to afford CP to reinforced concrete in different orientations.

In the 1980s, ICCP of reinforced concrete had also been used in a few countries around Europe and new types of anodes—principally, titanium meshes coated with mixed metal oxides (MMOs, which are a combination of ruthenium oxide and iridium oxide), outdoor paints dosed with graphite ground into platelets to maximise their electrical conductance and electrically conductive polymers – were being used on trials or in small installations which were carefully monitored. Certain of these early installations are still in operation today after passing current for more than 30 years. An example is in Switzerland on a railway bridge abutment at a location named Widerlager Brucke Rodi, Kt. Tessin. Here, 280 m² of concrete had an MMO-coated mesh and overlay applied in 1988, which is still operating effectively 30 years later. A conductive polymer anode wire was introduced by a large electrical company and aggressively marketed in both Europe and North America but was eventually discarded because of stress cracking of the plastic allowing the copper to be exposed, leading to anode failure. This type of system is shown in Figure 2.5.

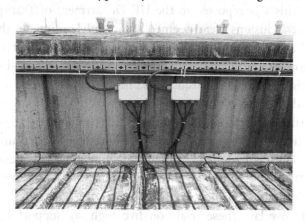

FIGURE 2.5 Conductive polymer wire before overlay added.

These new types of anodes were particularly relevant in Europe as bridge deck running lanes were protected by waterproofing, and it was the substructure of the bridges that was suffering widespread damage. Again in Europe as in the United States, it was the government road research authorities who had the money and knowledge to advance this remediation technology. The anode types of most interest in Europe, were those that could potentially be used in vertical and soffit applications, and the two American methods described above, which were in successful series use, were not even trialled.

ICCP had been used in Europe in isolated instances since the 1970s, but the first systemic consideration of the technique's veracity was by the UK Department of Transport that sent a representative (Kendell, 1985) to North America to interview all the leading agencies using this technique at that time in both the United States and Canada, and to inspect the most relevant installations. The main conclusions of this survey were that the Americans were happy that ICCP was established for bridge decks but considered that the technology had yet to develop sufficiently for substructures. The current density considered to be required for prevention of corrosion was in the range of 5–20 mA/m² of steel surface area. The galvanic anode systems were judged ineffective. This was reported to the UK Department of Transport, and its conclusions were deemed sufficiently optimistic that a parallel approach was followed with both a laboratory trial to determine its efficiency in stopping corrosion which had already started in reinforced corrosion and a field trial with four competing anode systems on motorway beams which had soffit, vertical and horizontal surfaces. The anodes in this trial were two types of conductive coatings (one a solvented chlorinated rubber and the other a solvented acrylic, both with graphite as their conductive element) – an experimental external ceramic tile arrangement as shown in Figure 2.6 and a conductive polymer wire in a sprayed concrete overlay. These trials on five highway support beams (one did not have CP) were heavily instrumented and run over

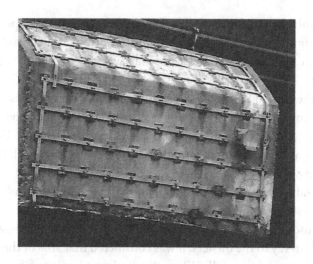

FIGURE 2.6   Experimental ceramic tile with titanium feeder strip.

several years. Later an MMO mesh with a sprayed overlay was added to the trial. Other laboratory trials were initiated to determine whether alkali aggregate reaction (AAR) could be triggered by ICCP, whether the bond between rebar and concrete could be softened, and finally the possibility of causing hydrogen embrittlement in susceptible steels. It was found in these trials that it was possible to stop active corrosion by imposing an external anode, but there were limitations on the anode's throw of current. With susceptible aggregates, it was possible to trigger AAR but the current levels required were much greater than that used in CP and similar to that used in electrochemical desalination, which is a chloride extraction process. Finally, it was found that only certain types of steel could be made to show any strength reduction, again at very negative potentials.

At the completion of the support beam trials, the English Department of Transport specified conductive coatings for commercial repair ICCP contracts for many years. After about a decade of fairly successful use, the Department of Transport specified that only water-based paints were to be used as anodes, and relatively untested anode products from North America were substituted for the solvent-based materials. These proved to be significantly less durable than the solvent-based products and have led to a decline in popularity for this type of anode system.

In the United States, as part of the Strategic Highway Research Program, a large-scale review into 'the CP of reinforced concrete bridge components' was undertaken (Broomfield et al., 1992). An initial survey covered a total of 287 CP installations. The vast majority of these systems were for decks, with the largest number conductive polymer in slots, followed by conductive polymer cable systems, and finally flat anodes in conductive asphalt. All of these anode systems are considered obsolete today. When they did a mail survey to the owners, it was considered that 90% of the systems were working well. When they independently verified this with site visits, it was found that several systems were switched off, there was anode damage, there were feeder wire problems and the success rate was considerably lower than what the owners had indicated. Particular problems were identified with conductive paint anodes, most particularly, current dumping in marine environments. The problem with the conductive polymer wire severing was noted. What was evident is that the original Stratfull-era anode systems were looking durable, but some of the newer generation of anodes was having significant problems, and this echoed the previous findings of Kendell. One of the principal recommendations of this report was that the clients provide funding for regular monitoring and maintenance of these systems, as this was not happening in the majority of cases.

In the mid-1980s, a widespread adoption of the principles of electrochemical corrosion prevention occurred in Scandinavia (particularly Norway and Denmark) where the Danes championed drilled-in or discrete anodes (with examples shown in Figure 2.7) and the Norwegians pioneered with desalination and realkalisation as shown in Figure 2.8. To this day, these nations have remained as world leaders in the application of these technologies to various reinforced concrete structures.

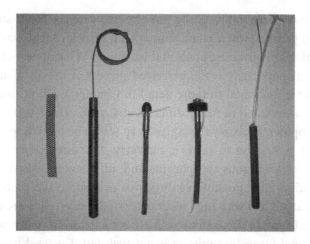

FIGURE 2.7 Various drilled-in anodes presently available. (Courtesy of John Broomfield.)

FIGURE 2.8 Spraying paper pulp as part of temporary anode for a real-kalisation system. (Courtesy of John Broomfield.)

After these early years, the use of CP has become more wide-spread. The growth has generally been incremental, with adoption for a particular problem in an individual country. Typically, its initial introduction has been on highway bridges, but this is not universally so. For example, in Germany, most of the structures which have been cathodically protected to date are multi-storey parking garages, whereas in Denmark, an early favourite was and is municipal swimming pools and balconies. What appears critical for its popularity is the perceived success of the installation, and this is discussed in more detail in Chapter 3. CP utilisation has been stretched to many different structures as it has become more popular. Probably the biggest leap was to be providing some form of protection to historic masonry structures, where steel girders, other ferrous components and, on a few occasions, lead are rusting after possibly centuries of exposure. The problems that Stratfull had with high resistances of the electrolyte are even more extreme in these stone or brick structures. In many cases, the normal protection criteria is not met, but it is used because it is the only practical repair technique that is possible on these often-listed structures (with example in Figure 2.9).

In the 1970s, some trials were done of electrochemical chloride removal (desalination) in the United States (Broomfield, 1997), but more fundamental studies and a large scale were not followed through until the late 1980s, by which time various temporary anode types had been developed. In some countries such as Switzerland, this technique was the most popular electrochemical repair technique up until this decade, but in most countries, in recent years, the use of desalination has become less popular while the use of CP has significantly increased. There are probably several reasons for this. One is that the high current required can make the concrete more porous. Another is that defining when success has been achieved is difficult. Finally, the chloride profile and reinforcement arrangement have to favour this procedure.

Electrochemical realkalisation is a similar process to desalination with the injection of alkali into the concrete by electromigration to restore the pH of the concrete. It was estimated that by 1994 (Broomfield, 1997), more than $30,000\,m^2$ of concrete surface area had been treated with a typical approach as shown in Figure 2.9, although since this time its popularity has waned. The reason for this is mainly that it has become apparent that the problem with

FIGURE 2.9 A department store in London with large-scale use of ICCP to prevent further damage to masonry.

carbonation can be treated with patch repair and coatings while this is not suitable for chloride-contaminated concrete.

In the 1990s, galvanic anodes were introduced initially to prevent incipient anodes in patch repairs. Due to their low cost and widespread marketing, these systems have become popular and are now marketed by some companies as suitable for achieving CP in all situations.

Since 2010, a hybrid system where galvanic zinc anodes are powered by an external power source for at least a week has been strongly marketed around the world.

The use of galvanic and hybrid anodes is now probably more widespread than ICCP as it tends to be used by a broader base of clients. The use of ICCP tends to be on larger and more complex projects where there is a more formal arrangement with an owner, consultant and contractor or on projects where degradation is most rapidly occurring and needs to be stopped. There is a division in the CP industry which is presently conflated on the merits of galvanic and hybrid anodes for cathodically protecting reinforced concrete structures, as in many instances, CP will not be provided but there may be some cathodic current and a reduction in corrosion rate.

There have been several ICCP installations for cathodic prevention on high-value projects in Europe, but it has not become a standard method for durability enhancement.

The Middle East has suffered from an extremely short practical life of reinforced structures particularly in coastal environments. This has been caused by poor construction practice, poor materials and, most importantly, an unusual aggressive environment. The damage caused has forced this region to be an early adopter of ICCP where it has been used on underground structures such as tunnels and chambers, bridges and many other projects. What made these projects notable was their large size and that the corrosion was previously occurring at a high rate. Additionally, the Middle East became the world's largest user of ICCP for cathodic prevention after it tried all the durability-enhancing alternatives such as epoxy-coated steel (not reliable), stainless steel reinforcement (uneconomic), inhibitors (did not work) along with increased concrete cover and innovative concrete mixes. In some of the Gulf countries with saline groundwater, ICCP for cathodic prevention is now mandatory for below ground reinforced concrete on government projects.

Japan and Australia have been in the forefront of the adoption of ICCP, with both countries heavily influenced by Scandinavian technology. Japan's structures have suffered severe corrosion damage for several reasons. Large amounts of infrastructure were built in the 1960s and are now approaching old age; Japan's mountainous nature meant that many of its highways and railways were placed in marine locations following the Pacific coast with high temperatures. The mitigating feature for Japan's infrastructure is that the bridges and other structures were designed to cope with earthquakes so they are structurally heavily engineered, so what is seen in Western eyes as extreme damage requiring demolition can be tolerated. Despite Japan being an early adopter of ICCP, it has become only sporadically used by the road authorities despite massive areas requiring treatment. Possibly, the huge cost of widespread adoption has caused regional road and rail authorities to defer projects and then replace the entire structure, as capital budgets are generally much greater than maintenance budgets. The significant rates of corrosion of these structures because of environmental aggressiveness have prevented the use of galvanic or hybrid systems.

Australia was an early adopter of ICCP, and it has become commonplace in all the coastal conurbations of the country as a repair technique. In recent years, impressed systems have been joined by hybrid and galvanic anodes in select applications.

In the United States and Canada, many of the systems installed after the original Stratfull era had poor design, installation procedures, materials and maintenance, and the use of ICCP for reinforced structures became less popular after the 1990s. This was illustrated by Sharp (2007), where the Virginia transportation authority undertook a survey of its nine installed CP systems. It read badly, as all the ICCP systems were abandoned after installation, with none of them operating when the survey was made. It is only with the positive influence from countries outside of North America that its popularity has risen again as a considered option for the repair of highway structures and buildings, though some of the recent installations have again been of poor quality. There is significant use of ICCP on bridge substructures presently in certain states, namely Oregon and Florida. The use of galvanic anode systems is widespread, and some regions such as Florida highways have published data on the use of these systems for marine columns for highway bridges. In this warm and salty environment, they found that this technique had variable success depending on the design of the jacket. ICCP has not been used for cathodic prevention in North America.

SUMMARY

Concrete is the most commonly used building material in the world, and nearly all engineered structures are combined with steel reinforcement to provide the optimum structural qualities along with durability. It was reported in 1909 (Gjorv, 2014) that some of the early reinforced structures were suffering from premature degradation, and by 1923, Atwood and Johnson (Gjorv, 2014) had assembled more than 3,000 references on the durability of concrete structures in marine environments, but it was not for another 40 odd years that electrochemical techniques were developed to combat this problem on reinforced concrete structures. These early systems worked well but were limited in that they had to be on the top of a bridge deck. In the last 35 years, many anodes have been introduced with markedly different designs. All have intended to pass current through the concrete, be durable and be economic. There have been significant failures in some of the anodes in many designs, and the situation coming up to date is that anode design particularly is still in an active state of development.

REFERENCES

Broomfield J, J Tinnea, *Cathodic Protection of Reinforced Concrete Bridge Components*, (1992), Transportation Research Board, Washington DC, USA, SHRP-C/UWP-92-618.

Broomfield J, *Corrosion of Steel in Concrete*, (1997), E & F N Spon, London, UK, ISBN 0419196307.

Gjorv O, *Durability Design of Concrete Structures in Severe Environments*, (2014), CRC Press, Boca Raton, Florida, ISBN 13-978-1-4665-8729-8.

Kendell K, *The Cathodic Protection of Reinforced Concrete*, (1985, May 13), Transport and Road Research Laboratory, UK, USA site visit.

Melchers RE, *Reinforced Concrete Corrosion Protection Repair and Durability*, (2017), Australasian Corrosion Association, Blackburn, Australia, ISBN 978-0-646-97456-9.

Sharp S, Survey of cathodic protection systems on Virginia bridges, Virginia Transport Research Council, VTRC 07-R35, (2007, June).

Stratfull R, Experimental Cathodic Protection of a Bridge Deck, Corrosion, 15(6), (1959), 65–68.

Stratfull R, *Experimental Cathodic Protection of a Bridge Deck*, (1974), Transportation Research Board, Washington DC, USA, Transportation research record 500.

Present Use of Impressed Current Cathodic Protection in Reinforced Concrete

AN ICCP SYSTEM

An impressed current cathodic protection (ICCP) system as shown in Figure 3.1 consists of three components with connecting wires. A power supply takes AC electrical power and delivers a low-voltage DC (typically 2 V to 8 V) to an anode system, which is distributed around the structure. This anode distributes and changes the electronic current flow into an ionic current through the concrete and onto the steel reinforcement. Sensors are placed in the structure to provide some measure of the system's functionality.

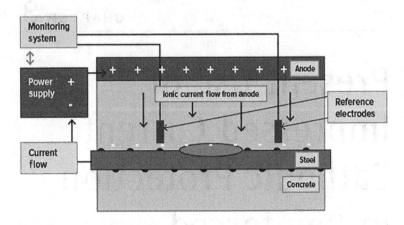

FIGURE 3.1 Schematic of an impressed current protection system for steel reinforced concrete. (Courtesy of John Broomfield.)

DEVELOPMENTS IN ICCP

Power Supplies

The first power supplies were taken straight from traditional cathodic protection (CP) practice and were tapped transformer rectifiers with a limited smoothing circuit. While these were extremely rugged and reliable, they were not suited to the low-voltage and current outputs of the smaller circuits required for the effective ICCP of reinforced concrete. The next development was to use thyristor control, which was in the early days basically a valve and is now similar to a transistor. These gave better control but were inefficient (about 60%–70%) causing heat problems, and the output required substantial smoothing. At the time electrolytic capacitors were used for the smoothing circuits, and they had a life expectancy or, as used in the electrical trade, a mean time to failure (MTF) of typically 8 years. Perhaps, because of ignorance, these units are still specified for reinforced concrete structures with data-logging equipment, which is not linked to switching of the output circuits.

Recently, switch-mode power supplies have become normal. Initially, they were secondary switchers (there was a step-down transformer at the intake of the circuit), but now they are all primary (that is they operate directly on the input AC voltage). This mirrors the use of switch-mode transformers, which are used for all modern electronic devices such as mobile phone chargers to electric car chargers. These systems work by first inverting a portion of the AC and then slicing the sinusoidal wave into thin slices (a typical duty cycle is 100 kHz for a modern unit), which are then brought back together through a smoothing circuit. These units have several advantages over the previous offerings. First, they are very efficient (can be better than 95%) so heat problems are greatly reduced; second, they marry well with electronic control circuitry; and third, with a ceramic smoothing circuit, they can have an MTF of more than 20 years for all the circuitry. On modern units, no moving parts such as cooling fans or disc storage would normally be allowed as these have a low MTF.

Modern power supply systems can now be remotely operated. This used to be by modem and is now commonly by wireless router through the Internet. Presently, there are four routes for this communication, namely fixed wire, asymmetric digital, subscriber line and wireless and global telephone. Some systems are now uploading data to the cloud, while others will send a monthly data bundle to a selected e-mail address. Recently, hacking and data infection have become significant problems, and some seasoned users are moving away from this remote operation. Another reason for this change in operating procedure is that they want to go to site to look at all of the CP systems and its interaction with the environment.

Early automated units were programmed in BASIC on a personal computer (PC). As the operating system called 'windows' became larger, the stability of these systems became more suspect. This was particularly important on these machines as they were not able to reboot without visiting the site to switch off and on the power, this was eventually 'solved' by an automatic reboot on a monthly basis. This solution while effective showed that there was a significant problem with the program base. In recent years, there has been a widespread adoption of a programmable logic controller (PLC) as the control computer, which runs a much more restricted control program. These PLCs are now almost universally used on all industrial control equipment. To make matters a little more confusing, the modern PLCs are virtual computers in a PC, an example is shown (Figure 3.2). Who said electronic engineers are rational!

There appears to be two approaches to modern power supply control system being offered for sale for the control and monitoring of a concrete ICCP system. The first is to adapt a system from an existing control set-up, typically this is a building control system. Building control systems are generally provided by large electronics companies for the specific task of automatically maintaining heat, humidity and light at optimum levels in peopled buildings. You then clip on modules, which are controlled by the PLC. This has been the most common procedure to date and has a strong advantage in that there is a low capital cost. The second approach is to purpose design electronics for the particular task of providing current and monitoring an ICCP. This approach allows the modules to be optimised for this task but has the disadvantage of requiring a substantial investment.

FIGURE 3.2   Touch screen control panel of a modern PLC power supply controller.

Power supply and monitoring electronics historically have been a significant cost component of an ICCP installation, but recent developments have allowed the price to fall to 20% of its level 15 years ago. This has meant that zone sizes have decreased and the number of reference electrodes has increased, sometimes to remarkable levels of overdesign. On a recent project, there were 62 sensors specified for 200 m² of ICCP.

Some of the things a power supply control and monitoring system ideally should do for the effective operation of an ICCP are to interrupt the current for all the output zones at the same time and then take 'instant off' values for each of the reference electrodes. This requires coordination of a timing system, switching off the output zones in a synchronized manner with a purpose designed storage protocol for simultaneous data capture which is all difficult. Consequently, many of the power supply systems offered do not actually do this.

About 50% of electronic problems seen on old ICCP systems are not caused by circuitry failure but by problems such as water ingress, vandalism, hinges collapsing or damage caused by lightning strikes and other transients. Great care should be exercised in placing the units in benign conditions as not seen in Figure 3.3, which can be easily accessed in the future.

FIGURE 3.3   Water damaged power supply enclosure caused by flooded basement.

Sensors

Many types of sensors have been used. The first and most direct are the current and voltage of the output circuit. When commissioning, this gives an indication of the combination of the anode circuit, the cementitious medium and the steel reinforcement. By happy coincidence, the current generally flows most easily where the corrosion circumstances are worst. Without this fact, it is likely that ICCP would not be practical. Changes in the output current and voltage will be expected for different weather conditions, the polarisation of the concrete, changes at the steel to concrete interface and damage to the anode.

In the past, several types of sensors have been tried. One of the more innovative was corrosion rate coupons. Cutting rebar in the native concrete was also attempted as a way of using the original steel and concrete as a coupon. In both of these, a zero resistance ammeter was introduced into the circuit and the idea was that the corrosion current measured without CP on these pieces of steel was gradually reversed as the CP was applied until the measuring coupon became a net cathode. The results obtained by this technique were difficult to analyse, and because the areas were very small, expensive high-resolution instruments were required, so this could not be part of the routine monitoring regime. This procedure has been quietly dropped.

Another technique, which is still marketed to this day, is electrical resistance coupons or wires. The idea here was that corrosion would reduce the area of the wire increasing the resistance. No useful data ever came out of these, and they are only rarely specified presently.

The most commonly used sensors are reference electrodes as they address the potential of the steel so the 'work' that the CP system applies to the steel can be measured. Typically, these are the only sensors used on a present day ICCP system. The first reference electrodes used followed traditional cathodic protection materials in the 1950s such as zinc and copper sulphate. Both of these had fatal flaws for the concrete application in that zinc was passivated and the copper sulphate was poisoned by the chloride ions that were in the concrete being measured. A system that was also popular was to leave 'windows' particularly with anode coatings so that a grid of the potentials could be taken around the structure, which could then be compared to the potential map taken from before the anode was applied. The biggest problem with this was that water or the gel used to electrically couple the reference electrode would often touch a part of the anode circuit giving strange readings. Various other reference electrodes were tried with an early favourite being silver chloride or more formally silver/silver chloride/potassium chloride as shown in Figure 3.4. This has worked well for most structures but does not seem to like aggressive marine conditions. A surprise winner in the reference electrode popularity stakes for reinforced concrete is manganese oxide. This is a little odd, as it is not based on a traditional fixed voltage difference between an element and its ion. Its difference in potential is caused by the differing voltage of the different oxidation states of the manganese ion which is related to the output of

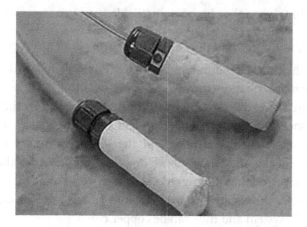

FIGURE 3.4   Silver/silver chloride/potassium chloride reference electrodes for embedment in reinforced concrete.

an alkaline battery cell. These reference electrodes were successfully prototyped, and they have proven to be the most stable cell so far developed, particularly in aggressive conditions, with more than 25 years track record.

There is also a category of reference electrodes called pseudo reference electrodes which are usually inert material. In the past, graphite was commonly used but now the most popular is mixed metal oxide (MMO; this is a mixture of transition metal oxides) on titanium. The reason for its popularity is the extremely long theoretical lives being required by some of the clients and the operation of a measuring circuit on a frequent basis with a modern automated control system. For example, with a silver chloride reference electrode you can calculate a consumption time of the silver element of less than 20 years with a 20-min monitoring protocol. This time period may be unacceptably short for some specifiers.

On some modern electronics systems, a health check on the reference electrodes is undertaken which takes not just the potential but also the impedance. This has proven to be a valuable additional reading, as it allows removal of these cells from the control algorithms when they become unreliable.

Connections and Junction Boxes

When looking at the durability of an ICCP system, it is striking as to how many are inoperative due to poor detailing of the wiring.

In an aggressive environment, everything buried in the concrete should be of a conductive non-corroding metal, which includes reference electrodes and the positive primary anode feeders. Ideally, they should both be of titanium, but for the reference electrodes, a high-grade stainless steel-stranded core is acceptable. The negative connections to the steel are protected by the ICCP system and thus can be copper cored.

The practice of making a site connection with a bit of heat shrink between a copper-cored wire and a titanium anode feeder, which is going to buried in the concrete, should never be used.

Early systems used polyvinyl chloride sheathed cables. These seemed to be reasonably durable in a concrete environment and out of it, but they are not normally specified presently because of their smoke problem if ignited. Experience with cross-linked polyethylene and its American cousin high-density polyethylene (HDPE) has been reasonably good, but bad experiences have occurred when cable makers of dubious ancestry have been used. In the past, fluorinated ethylene propylene and silicone sheathed wiring have been successfully used. In all cases, the wiring should be double insulated.

Junction boxes should never be placed in areas where there is significant water ingress and should always be filled with a purpose-designed re-enterable gel resin. The practice of putting positive and negative connections in the same junction box should be deplored unless the location of the junction box is permanently bone dry.

Anodes

There are several categories of anodes that are currently in use, namely:

Organic Coatings

These were external paints, which had significant amounts of carbon black added to make them electrically conductive. The balancing act was to add sufficient carbon black (fine carbon used for making car tyres) for the electrical conductivity while still retaining the bond of the binder. The most promising paint tested had a chlorinated rubber extender, which was acid resistant. The highways agency in the United Kingdom adopted this and similar paints which were solvented. These were then used on the motorway bridges, as they were relatively cheap and maintainable. After their early successful use, the Highways Agency decided, for environmental reasons, to change all paint systems to water-based coatings including anode paint systems. The water-based anode systems subsequently installed were from Canada and America, and untested in the United Kingdom, they failed after a short time. This has severely dented the popularity of this type of anode. In a survey of 100 ICCP systems (Polder, 2012) in Holland, the Dutch Road Authority had looked at many paint systems and found that they deteriorated and required repair particularly where there was water leakage but they had protected from corrosion adequately (Figure 3.5).

Recently some new versions of conductive paints, which seem to offer good adhesion, conductivity and are durable have been introduced, and these are making some headway in Europe.

Metallic Coatings

Various metal coatings have been applied to the concrete over many years. Some of these have relied on the reactiveness of the metal to provide sacrificial cathodic protection and these are not considered further as this section is on ICCP systems. Various flame- or arc-sprayed metals such as stainless steel or titanium have been applied (Figure 3.6). Most have been moderately effective when first energised but have deteriorated in performance as the oxide product increases in thickness at the anode to concrete interface. An installation tried in the United States was of a

FIGURE 3.5 Damage to a conductive coating anode.

FIGURE 3.6   Flame sprayed titanium anode applied in Japan.

mild steel anode (Kendell, 1985) sprayed on the concrete surface (unsurprisingly it rapidly corroded away). Very few installations have been made in the last 10 years.

Primary Anode Conductors
Several primary anode conductors have been used with various systems, and they are discussed separately below. What is meant by primary anode is that the anode reaction occurs at the surface of this material.

Some of the early primary anodes were silicon iron, which was quickly dropped.

A more enduring primary anode with a dismal record is carbon fibre. Recently, another anode system with a carbon fibre primary anode was announced. One of the biggest problems with carbon is making a durable connection. There are others.

A popular American primary anode was a copper core with a niobium sheath under platinum. This worked well if it was not mechanically damaged during installation but was more expensive than the alternatives and was only available in rods and not flat strip.

Another American primary anode was an electrically conductive HDPE sheath over a copper core. This stress corrosion cracked after a few years causing the anode to fail.

A British invention (Hayfield, 2002) was a conductive form of titanium oxide (Ti_4O_7), which was trialled as conductive tiles but is now more commonly sold as tubes which are placed in drilled holes in the concrete. This has an advantage that it is more resistant to hydrogen fluoride but is difficult to connect to, is fragile, has a lower conductivity and may revert to its previous non-conductive oxide state (Sergi, 2007) after 10 years of passing current.

Today, the overwhelmingly most popular primary anode is MMO on a titanium substrate. This is a conductive ceramic of iridium and tantalum oxides, which is sprayed onto the substrate and then baked. This has virtually no flaws in that you can connect to it easily, the corrosion rate at the output level for concrete ICCP is practically zero (it is likely to have a far longer life than that predicted by the American Society for Testing and Materials rapid testing, as this assumes that there is a linear relationship between the current passed and consumption, which is almost never the case), and it is mechanically tough. Its only minor disadvantage is that the conductivity of titanium is fairly poor. This can be easily designed around with a little care. The other point that should be noted is that the titanium substrate has a breakdown potential of 8 V; this means that above this voltage, there can be significant degradation of titanium. In practice, this is not significant as the anodes should have an operating voltage way below this. In all studies, for example, Polder (2012), this anode has survived unscathed for more than 20 years. A similar experience was reported from Australia (Cheaitani 2017). The only failures have come from Chinese copies.

Cementitious Coatings with Additives

These systems have electronically conducting additives for improved distribution in a special mortar mix. The additive was carbon-chopped carbon filaments coated with nickel. When first launched, they were suggesting primary anode feeders at 1 m or above spacing. In the last installations, the spacing was down to 300 mm. Despite proving to be a durable anode, this system had faded in popularity probably due to cost but is making a minor comeback.

Slotted Anodes

In area terms, this is now the most popular of the anode systems. Most of the systems installed have used expanded MMO ribbon, but recently, there has been a significant upswing in using solid ribbon with MMO due to its higher current output and easier installation. In general, these systems have performed well, but in aggressive environments, current shunting has meant premature failure of the cementitious grout, typically within 5 years. What is happening is that any small voids between the ribbon and the concrete are getting filled with freshwater or, worse, salty water. This area then passes a lot more current than the rest of the ribbon, which causes more anodic product, which is acidic. This acid attacks the cementitious mortar increasing the size of the void until the primary anode is totally disconnected from the parent concrete (Figure 3.7). The reason for this anode's popularity is its low cost and no change in structural dimensions.

FIGURE 3.7   Slotted anode failure on a marine structure.

Anode Ribbon before Casting
This is by far the most common way of installing ICCP for new structures. The MMO-coated titanium ribbon is separated from the steel by an arrangement of plastic spacers. Typically, the spacing has been between 300 and 500 mm. This has worked well where sufficient detailing has been applied so that the ribbon is kept electrically separated during the casting process. One or two other solutions for anodes in new structures have been tried with mixed results.

Anode Mesh in Cementitious Overlay
This uses an expanded MMO on titanium substrate mesh as an overlay over the structure, which is then covered with a cementitious overlay (Figure 3.8). This overlay may be cast or sprayed on. Generally, experience with this anode has been good. The chief problem has been delamination of the overlay, which typically occurs because of insufficient surface preparation of the substrate. Some early observers linked this delamination to the passage of current and this is possible. Surprisingly, even when there is significant delamination, the ICCP keeps working.

FIGURE 3.8 MMO mesh before cementitious overlay applied in a Dresden car park.

FIGURE 3.9 Discrete anodes being installed inside a segment of the Saint-Cloud Bridge, France.

Discrete Anodes

These are anodes placed in drilled holes around the structure (Figure 3.9). These have become popular worldwide and are now widely used in many of the ICCP installations where they have proven to be durable and quick to install. The more sophisticated examples of these have an individual resistor built into each anode. This has significant benefits in better current distribution. Special versions have been developed for use in soffit applications. The biggest problem with these high technology anodes is cost, so some low technology alternatives have been used.

STATE OF THE ART

Sustainability

The residual life of a corroding reinforced concrete structure can be significantly improved by ICCP. The jury is, in my opinion, still out on sacrificial anodes delivering anything like the same improvement, and the use of inhibitors has proven to be disappointing. The only other method of increasing the residual life in the same environment is desalination, and this has proven expensive and generally been abandoned. Of course, this list does not include the removal of the chloride-contaminated concrete, which can be impractical with the chloride level required for ongoing corrosion prevention also not easily determinable. So in practical terms at present, ICCP is the only option if the owner wants to significantly extend the life of a structure without a dramatic change in the environment.

The most expensive part of the ICCP is the anode and for this to make an optimal repair has to be durable. Fortunately, there is now significant long-term experience with CP systems, which can guide us in what is the most suitable for a particular structure. Several countries have looked at their stock of ICCP projects and looked at how durable they have been. One of these was by a US Government research program (Sharp, 2007), which showed disappointing results with all of the ICCP systems being inoperative. In Europe, the Dutch (Polder, 2012) looked at the relative life expectancy of 100 ICCP systems and found that those with MMO-coated titanium primary anodes had generally survived for up to 20 years, and those with conductive coatings had done less well. Italy, which applied large quantities of ICCP in the 1980s, found that none of them had been wired to operate.

The difference between Northern Europe and the United States on the survivability of ICCP systems is large and probably reflects the cultural differences found in construction. This is that the life expectancy and quality of construction is significantly higher in Northern Europe, and the clients are more willing to pay up for maintenance.

ICCP systems are active, and thus there is a requirement for maintenance, even if this is as little as checking that the system is passing current on a quarterly basis. If there is no mechanism and expertise for this maintenance, then ICCP should not be selected as an appropriate repair option.

It is striking that large companies in a recent conference (Huang, 2017), who should know better, are still researching corrosion inhibitors as the panacea for the corrosion evil despite its proven failure to do any good whatsoever for fundamental reasons. One can surmise that the only reason for this is the marketability of a one-stop, fit-and-forget solution to the corrosion of steel reinforcement.

One aspect of sustainability, which hitherto has been ignored, is the circuit resistance of the anode. If the resistance is higher, say it takes 5 V to pass a certain current rather than 2.5 V, and then roughly, the electricity used will be doubled. This could be a significant waste of resource over the life of the structure. This factor and the potential longer life of the anode at lower current outputs means that anode spacings and maximum current output requirement should be conservatively configured.

CONCLUSIONS

1. ICCP does work and is effective in preventing further corrosion of steel reinforced concrete.

2. There are not any practical alternatives at present to CP for the rehabilitation of chloride-damaged reinforced concrete structures.

3. The success and failure of ICCP as a technique has been dramatically influenced by the design, construction and maintenance standards of a particular country.

4. For an ICCP system to be selected as a suitable repair technique, there must be confidence that the system must be adequately monitored and maintained for its design life.

5. The various items comprising an ICCP system have now all been tested in commercial installations so that realistic life expectancies can be predicted and achieved.

6. The proven quality and performance of demonstrably effective products should be drawn upon when designing and specifying ICCP systems.

7. There is no theoretical level of protection current, and the best course of action is to pass as much current as possible, in an even manner to the reinforcement, for as long a period of time as possible.

REFERENCES

Cheaitani A, *Review of Cathodic Protection Systems for Concrete Structures in Australia, (2017)*, NACE: New Orleans, USA, 9024-SG.

Hayfield P, *Development of a New Material–Monolithic Ti_4O_7 Ebonex Ceramic*, (2002), Royal Society of Chemistry: Cambridge, UK, ISBN 0-85404-984-3.

Huang E, *Considerations for Concrete Corrosion Control Alternative*, (2017), NACE: New Orleans, USA, 9046-SG.

Kendell K, *The Cathodic Protection of Reinforced Concrete*, (1985, May 13), Transport and Road Research Laboratory, Bracknell, UK, USA site visit.

Polder R, *Performance and Working Life of Cathodic Protection Systems for Concrete Structures*, (2012), P157 Concrete Solutions, Taylor and Francis, London, UK, ISBN 978 0415 616 225.

Sharp S, Survey of cathodic protection systems on Virginia bridges, Virginia Transport Research Council, USA, VTRC 07-R35, (2007, June).

Sergi G, *Long Term Behaviour of Ceramic Tubular Shaped Anodes for Cathodic Protection Applications*, (2007 October), NACE conference, NACE: Tennessee, USA.

Stratfull R, *Corrosion*, NACE International: Houston, USA, 15(6), (1959), p. 65–68.

CHAPTER 4

Present Use of Galvanic Anodes for Cathodic Protection in Reinforced Concrete

HISTORY

The history of using certain metals to galvanically protect other metals from corrosion by Sir Humphrey Davy and his pupil Michael Faraday is well known. Here, the aim was to protect ships in salt water. As time progressed protecting steel in the ground or water has become routine using galvanic anodes, and three galvanic anode material alloys have become dominant. These are zinc, aluminium and magnesium. These materials are typically very pure or have alloying additions. In all these types, the main objective of these additions or the high purity is to prevent the formation of an oxide film which is stable enough to hamper the dissolution of the metal, and hence reduce its current output.

Magnesium is normally modified with either manganese or with aluminium and zinc to prevent adherent film formation. The magnesium anodes are commonly placed in bags of a rapid wetting backfill comprised of gypsum, granular bentonite and sodium sulphate. The high potential of this metal means that magnesium anodes can be used in soil and freshwater applications. It normally corrodes too fast to be used in saline applications with the active life limited in this environment, but some American companies actively promote these anodes' use in salt water, particularly for descaling operations or where high currents are required.

Aluminium is normally modified with indium (in the past, mercury was used but its toxicity has caused it to be abandoned for environmental reasons). Aluminium anodes can be used in sea water and the sea bed. It is not normally used in freshwater or soil as it will passivate. BS EN 12496 (2014) limits the use of aluminium alloy anodes to brackish waters of resistivity less than 200 Ohm cm. Aluminium anodes have a high electrochemical capacity, so they provide a higher current output per kilo than zinc. They were more expensive, but this is changing, as zinc ore supplies are limited and more aluminium smelters are operating causing a reduction in world prices, they have comparable costs per kilo presently. This alloy is widely used in the offshore industry for the cathodic protection (CP) of structures and marine pipelines.

Zinc anode materials are either extremely pure or alloyed with elements such as cadmium and aluminium in order to prevent an adherent corrosion product building up. It is used in sea water and in low-resistivity soils, typically with a resistance of less than 1,000 Ohm cm.

All these alloys have been successfully used in the ground or in water to cathodically protect steel in concrete. In both these cases, the large electrolyte envelops both the anode and the reinforced concrete structure. These structures are, for example, a reinforced concrete tunnel under an estuary with the anodes mounted either on the surface of the concrete or in groundbeds. These types of installations are in common use around the world, and the traditional criteria used for determining the success of these CP systems is being used unaltered.

Galvanic anodes were trialled in the 1960s by the US Highways Administration where the anodes were placed directly in the concrete to protect the steel reinforcement, which was actively corroding. They concluded that the galvanic anodes did not work (Kendell, 1985) as they did not pass enough current and concentrated on developing impressed current systems.

A problem with patch repairs in chloride-contaminated concrete structures causing further corrosion was noted, and solutions were researched. The 'incipient anode effect' is one where a patch repair with clean mortar causes new corrosion problems on the steel in contaminated concrete surrounding the patch (see Figure 1.5). One way of stopping this was to incorporate galvanic anodes into the patch repair, and another was to apply a coating to the exposed rebar in the patch. Because of the well-known problems of zinc anodes passivating, some of the anodes that were used had a cast mortar surround with the addition of lithium, nitrite or chloride compounds to the mix.

Since this research was undertaken (Sergi, 2008), many commercial galvanic anode systems have been brought to the market for use directly in or on concrete. First, they were supposed to prevent corrosion from occurring, cathodic prevention, and secondly, some of the systems being offered are stated to be in full compliance (EN 12696, 2012) for the CP of reinforced concrete, Compliance with this standard has the aim of preventing ongoing corrosion from continuing. These anodes are cheaper to buy, are typically cheaper to install and, as some cannot be assessed, have zero cost for maintenance relative to an impressed current system. Because of these advantages, their use is increasing significantly. The most pertinent question around these galvanic anode systems is to what extent are they effective?

PRESENT SITUATION

There are many systems that are presently being offered. These can be split into five categories.

1. *Surface layer*: In the United States, thermally sprayed zinc is used. In the United Kingdom, a thermally sprayed aluminium with small amounts of zinc and indium is offered. For these systems, there is sometimes some surface preparation required, but no chemical modification of the concrete surface to reduce the resistance. In the past in the United States, sheets of zinc foil that are glued to the concrete have been used. This has been abandoned in the United States mainly due to disbondment issues but continues to be in significant use around Europe. The hydrogel adhesive is also designed to disperse the corrosion products of zinc (Figure 4.1). A zinc-rich paint has also been trialled. These surface layer systems have a high surface area of 1 m² of zinc surface per m² of concrete.

2. *Anodes cast into a cementitious material*: This mortar normally contains a chemical ingredient such as lithium, nitrites or chlorides to prevent passivation of zinc. These anodes are

FIGURE 4.1   Zinc foil with gel adhesive. (Courtesy of John Broomfield.)

then placed in broken out areas or drilled holes. Some of these have several discs of zinc or some other arrangement to maximise the surface area. Note, in the example shown in Figure 4.2, there is a connection detail to allow direct connection to the steel reinforcement. Taking data from reported installations and data sheets, these systems have a surface area of approximately 0.1 m^2 of zinc surface per m^2 of concrete.

3. *Rods with or without a coating*: The anode coating or paste is to prevent passivation of zinc (see Figure 4.3). The anodes are placed in drilled holes and grouted with a cementitious material. Some versions of this system use an external power supply to drive these anodes for a limited time period (nominally 2 weeks), and these have been dubbed a 'hybrid system'. Taking data from reported installations and data sheets, these systems have a surface area of approximately 0.1 m^2 of zinc surface per m^2 of concrete.

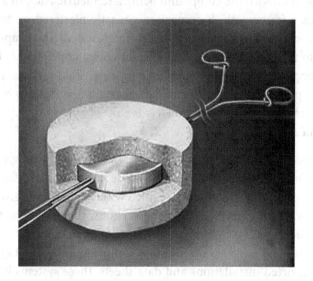

FIGURE 4.2 Zinc disc cast into mortar with attaching wires to reinforcement. Sacrificial with mortar puck. (Courtesy of Vector.)

FIGURE 4.3   Holes for zinc rods with 'daisy chain' connection slot.

4. Zinc mesh encapsulated in an alumo-silicate (AlSiO$_4$) compound to prevent passivation (refer to Figure 4.4). This is possibly the most sophisticated galvanic anode system seen to date with the compound being a tetraedric AlO$_4$ in a silicate framework. It comprises a mastic plastered on to the concrete surface. Then a layer of perforated zinc is applied to this, and it is encapsulated. An epoxy upper coat is then applied with a polyurethane topcoat. Taking data from reported installations and data sheets, these systems have a surface area of approximately 0.8 m^2 of zinc surface per m^2 of concrete.

5. *Jackets for marine applications*: These typically comprise a fibreglass jacket with the zinc anode normally in the form of a mesh, with the void space filled with cementitious grout as shown in Figure 4.5. In some versions, this grout has been loaded with a higher cement content to maximise the pH, but most use a standard flowable mortar. Taking data from reported installations and data sheets, these systems have a surface area of approximately 0.8 m^2 of zinc surface per m^2 of concrete.

Embedded zinc activating (EZA) - solid matrix serving as electrolyte

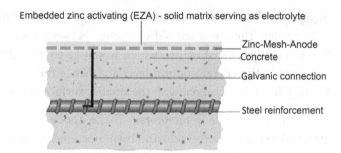

Zinc-Mesh-Anode
Concrete
Galvanic connection
Steel reinforcement

FIGURE 4.4 Alumo-silicate-based anode. (Courtesy of CAS-composites.)

FIGURE 4.5 Marine jackets. (Courtesy of Structural Group.)

All of the commercial systems presently on offer have used 99.99% pure zinc apart from the sprayed aluminium offering.

There are three fundamental differences between the commercial offerings:

1. *De-passivation of the zinc*: Some of the systems offered here have no way to reduce the resistance of the anode circuit. The method of de-passivation needs to be effective over the life of the anode.

2. *Significant differences in the respective surface areas of the zinc anodes*: There is a tenfold difference between some of the anode types listed earlier, which would give immediate concerns at the current output level and hence the level of CP which can be achieved with the smaller surface areas.

3. *Ability to be externally monitored*: If the anode is simply attached to the reinforcement and then mortared in, its output current and voltage cannot be monitored. In this case, this anode system is not in compliance with the code of practice (EN 12696, 2012) for the CP of reinforced concrete. It is interesting to note, certain manufacturers are plainly showing a direct connection detail for their anode while stating on the same page it is in compliance with EN 12696.

HOW GALVANIC ANODES WORK

Zinc has a natural potential of $-1.05\,V$ (with respect to silver/silver chloride electrode/$0.5\,M$ potassium chloride), which is significantly more negative than the reinforcing steel, typically $-0.35\,V$ to $-0.5\,V$ it is trying to protect. This difference in potential is the driving voltage. As current is passed, the zinc metal will be consumed and corrosion products will be formed on its surface. This is exactly the same process as in a car battery or a zinc battery, but in these two examples, the electrolyte is sulphuric acid or ammonium chloride which are both designed to minimise the build-up of corrosion product on the zinc anode's surface. Build-up of corrosion products will reduce the driving voltage and current.

The consumption of the zinc to form ions generates electrons that flow from the anode, through the anode connection to the steel. The build-up of electrons on the steel depresses the potential over the whole surface of the steel making it all a cathode and eliminating any anodic reactions on the steel. Excess electrons leave the steel by a process known as charge transfer, where water and oxygen combine to form negatively charged hydroxyl ions. The hydroxyl ions increase the pH, passivating the steel.

Zinc used for anodes is selected so that the corrosion product does not completely passivate its active surface through the formation of hydroxides or other compounds. The zinc has 99.99% purity to prevent iron impurities stabilising the corrosion product formed. In warm, fresh and stagnant water, even this anode will passivate and then provide a negligible current output. This is why in conventional CP (not concrete) installations, zinc anodes are generally not used in high-resistivity environments.

Zinc metal is normally cast around a mild steel or galvanised support called an insert, which is continued outside the anode. Other supports can be stainless steel. This support is then directly connected to the steel reinforcement or attached to an insulated or uninsulated wire, which can be run as a string with many anodes attached, brought to the surface of the structure and electrically connected through a junction box to the steel reinforcement. This latter arrangement has the advantage that it can be attached through a switchable connection, so current flow can be measured with a zero resistance ammeter and current interrupted to get a polarisation resistance-free measurement from sensors measuring the potential of the steel reinforcement, commonly known as reference electrodes. This arrangement also allows depolarisation testing which is required in EN 12696.

To help reduce the grounding resistance, restrict the formation of passivating films and prevent electro-osmotic dehydration, it is common to use backfill compounds to surround the galvanic anode in soil applications. Typically, this could be a combination of gypsum, bentonite and sodium sulphate with the anode encapsulated within a cotton sack containing this compound or a bore hole dug and the anodes interspersed with the backfill. In flowing sea water, this is not necessary, as the water flow sloughs the corrosion product from the zinc.

LIMITATIONS

The driving voltage of zinc is limited to a maximum of 0.7 V with typical potentials obtained from actively corroding steel in concrete, whereas an impressed current CP (ICCP) anode can be 5 V or more, so with an impressed anode of the same size, you should be able to roughly output seven times more current. This is why when the first galvanic anodes were being introduced for concrete (Sergi, 2008), they were for preventing 'incipient anodes' and not designed to prevent ongoing corrosion. This distinction seems to have been lost on some of the latest galvanic anodes brought to market (Mapei, 2012), which recommends their product for use in protecting against corrosion in areas of the structure requiring repair work; in other words, places where active corrosion is occurring. The placement of these anodes also appears at odds with accepted practice for impressed current anodes with a spacing of up to 800 mm centres shown, whereas an ICCP discrete anode system will typically be spaced at a maximum of 400 mm centres despite its much greater current output. So using the Mapei worst case, a quarter of the anode numbers (400 mm to 800 mm spacings) with a seventh of the likely output at each anode means the output current in a given square metre is likely to be 3.6% of an impressed system. This makes it very doubtful that the normal criteria of success with which CP of reinforced concrete is normally measured will be achieved.

When a galvanic anode is in sea water, as zinc reacts, the corrosion product dissipates into the solution, which continues to intimately surround the zinc despite its change in shape. Typically, the utilisation factor (the ratio between the theoretical output and the actual output) is between 0.7 and 0.9 for this sea water environment depending on the way the anode has been shaped to promote other parameters like output and current distribution. For zinc anodes cast into a mortar mould or grouted directly into a hole, the utilisation factor is likely to be dramatically lower because as the zinc corrodes, it will shrink, and then there will be either voidage or it will be filled with the non-conducting zinc corrosion product. These voids, unless filled with water or pore solution, will reduce the output of the anode dramatically.

Data were provided by Sergi (2008), who monitored galvanic anodes at 13 locations in concrete in the first year, second year and subsequently. He reported there was a significant difference in behaviour. With seven sites, the outputs diminished by more than 50% and the most extreme dropping by 97%. In four of the sites, the output appeared to be relatively stable, with the other two results in the middle. These very different outcomes can possibly be explained by some of the galvanic anodes passivating and others not.

Some factors for how well a galvanic anode will actually perform are discussed below.

TEMPERATURE

Zinc is a relatively reliable anode, but it is temperature sensitive for its output. In an output capacity against temperature relationship in saline mud, the pure zinc anode output drops to zero below 15°C (Shrier, 1994) and again drops to zero above 50°C, at which point it becomes cathodic to steel depending on the environment and the exact composition of the anode material. It is advised that above 40°C, pure zinc will become passivated, whatever its environment (Shrier, 2010). In further research (Slunder, 1983), this potential reversal is dependent on the chemical environment surrounding the anode. If there is gypsum surrounding the anode, then it is unlikely to occur; unfortunately, this compound is unlikely to feature in a reinforced concrete galvanic anode circuit.

Thus, this output change with temperature can be expected to occur in concrete, mortar or other relatively high-resistance environments. The reason for these changes is probably that at temperatures where there is maximum corrosion of the anode, and hence maximum current output, the corrosion film on the surface of the anode is granular and non-adherent. At lower temperatures, it is gelatinous and adherent, and at higher temperatures, it is compact and adherent (Slunder, 1971). What this means is that in Northern Europe throughout all the winter months, a zinc galvanic system will not provide much or any CP. Plainly, the claims of some of the manufacturers (Mapei, 2010) that their anodes service range is from −10°C to +80°C is nonsensical.

What is not commonly recognised is that the design current density guidelines (DNV, 2010) for cathodically protecting bare steel in sea water increases as the temperature reduces. Four categories of temperature are categorised, namely tropical, sub-tropical, temperate and arctic. While this is for both galvanic and impressed current systems, it may be a partial reflection on the reducing output at lower temperatures of galvanic anodes. Single reaction chemistry models where the reaction rate increases with increasing temperature is in direct conflict with these DNV guidelines which are based on practical experience.

CHLORIDE LEVEL

Zinc is not recommended for use in freshwater but is recommended for use in salt water. This suggests that sodium chloride affects the corrosion reaction, the corrosion product or possibly both. Generally, the highest resistance environment zinc anodes are recommended for use is 1,000 Ohm cm. This is an unusually low resistance for even heavily chloride-contaminated and aged concrete. If the galvanic anodes are driving through chloride-contaminated, wet and warm concrete, it should be easier to pass an adequate amount of current. Against this is the possibility that the corrosion reaction occurring on steel in this circumstance will require more electrons to negate its effect, and thus the anode will have to pass more current to be effective. Zinc is not directly activated in the same way as aluminium anodes, and the critical factor determining its output appears to be the maintenance of a porous corrosion film, which allows the passage of oxygen.

ANODE DESIGN

The categories of anode design and then the number of individual anode systems commercially available within each category surprised the author when researching this section and would seem indicative of the sales success of the use of galvanic anodes in cathodic current application (electrochemical treatment). Some of the designs appeared doomed to failure while other designs were more sophisticated and possibly could provide some form of corrosion prevention or corrosion reduction benefit to steel in concrete.

AMOUNT OF ZINC

Most of the life expectancies of the anodes are formulated using the calculation which is used for consumption of galvanic anodes in sea water. This has an amp-hour capacity for the zinc inserted along with a utilisation factor and a current density to give a life in years. This equation is not valid for zinc in concrete as the current level is likely to be severely restricted by the anode circuit resistance and the utilisation factor is likely to be hugely influenced by the anti-passivation measures taken and the immediate environment around the anode system, for example, if the surrounding concrete is dry. In some instances, it has been attempted to take anode resistance formula for various shapes of anode using equations such as the Modified Dwight, Lloyds and McCoy to identify the likely output of zinc anodes in concrete. These equations pertain to long slender anodes, thin plates and bracelet anodes respectively, and all were empirically derived from practical experimentation of anodes in salt water baths with an uncoated steel cathode. This circuit arrangement is fundamentally different to the actual protection circuit arrangement of a zinc anode driving through concrete onto reinforcement.

REFERENCES

BS EN 12496, *Galvanic Anodes for Cathodic Protection in Seawater and Saline Mud*, (2013), BSI, Brussels, Belgium.

DNV-RP-B401, *Cathodic Protection Design-Rules and Standards*, (2017), DNV-GL, Oslo, Norway.

ISO 12696, *Cathodic Protection of Steel in Concrete*, (2012), British Standards Institute, London, UK.

Kendell K, *The Cathodic Protection of Reinforced Concrete*, (1985, May 13), Transport and Road Research Laboratory, UK, USA site visit.

Mapei, Mapeshield, data sheet 6102, (2012, July).

Mapei, Galvanic cathodic protection brochure, (2010, February), p. 27.

Sharp S, Survey of cathodic protection systems on Virginia bridges, Virginia Transport Research Council, USA, VTRC 07-R35, (2007, June).

Sergi S, D Simpson, J Potter, *Long Term Performance and Versatility of Sacrificial Anodes for Control of Reinforcement Corrosion*, (2008), ProcEurocorr 2008, European Corrosion Congress, Edinburgh, UK.

Shrier L et al., *Corrosion*, (1994), Butterworth Heinemann, Oxford, UK, vol. 40, p. 10, ISBN 0750610778.

Shrier L et al., *Shreir's Corrosion*, (2010), Elsevier, Holland, vol. 4, no. 19, p. 2763, ISBN 9780444527882.

Slunder C, WK Boyd, *Zinc: Its Corrosion Resistance*, (1971), Zinc Development Association, London, UK, p. 102.

Slunder C, WK Boyd, *Zinc: Its Corrosion Resistance*, (1983), Zinc Development Association, International Lead Zinc Research Organisation, New York, USA, p. 223.

CHAPTER 5

Future Use of Cathodic Protection in Reinforced Concrete

INTRODUCTION

Cathodic protection (CP) in both its forms of impressed and galvanic is becoming used in more countries and in greater quantities. The future of CP is bright if the systems presently being installed do provide long-term performance; that is, they keep passing current evenly through the concrete for a long time to the protection criteria of an applicable standard. In certain areas, such as North America, this level of performance has not been reached, and in other countries, there is an uneven set of performances, so one of the most important aspects in the future success of CP is that appropriate CP installations are made in the future, and they are demonstrated to be working in accordance with relevant standards.

For this validation to occur, thought must be given on how to monitor in a way that can more directly and accurately demonstrate that the application of cathodic current prevents further corrosion. This monitoring is particularly necessary for galvanic installations as the present situation is that the majority of installations lack any form of monitoring and also uncertainty that they are providing the benefit that was hoped for.

USE IN NEW STRUCTURES

The uptake of impressed current cathodic prevention as a life enhancement technique has been limited outside the Middle East where structures such as the Jeddah Tower, Saudi Arabia, which will be the world's tallest building when completed, is having its foundations protected by CP. This tower is shown in Figure 5.1. Until recently, the large marine tunnels in North America and Europe have not used this technology, but it is being actively considered for some of the latest projects.

One of the biggest problems in preventing CP being more widely used is that on large structures, unrealistic theoretical models about the actual lifetime before corrosion damage appears are being routinely made (Figure 5.2). One example of this is a large marine bridge in Northern Europe with very-high-strength concrete (90 MPa), which was estimated to have a minimum life expectancy of more than 100 years before there was any corrosion damage. In reality, it required a repair impressed CP system being installed within 10 years of construction because no account was taken of the poor ductility of the concrete that had been modified by silica fume additives.

Presently, the most common ICCP anode for cathodic prevention is mixed metal oxide (MMO)-coated titanium ribbon spaced off the cage (Figure 5.3), but there are some suggestions and have been large-scale trials of alternative anode systems. Galvanic anodes are incorrectly being marketed for this application due to the high resistance of new concrete.

FIGURE 5.1 The world's tallest building in construction, the Jeddah Tower, has cathodic prevention anodes protecting its foundations.

FIGURE 5.2 Bridge with premature corrosion damage.

One of the more unfortunate parts of cathodic prevention is that even though it has been specified on several large-scale structures, when the contractor puts his total bid which is commonly over the budget, the CP is then dropped. This happens in more

FIGURE 5.3 MMO-coated ribbon spaced off the steel reinforcement. (Courtesy of CPT.)

than 50% of all new projects where cathodic prevention is specified. Probably this is because the expense of installing this system can be deleted with no change in the short-term operability of the structure. As with most durability issues, there is a recurring theme of upfront cost against predicted life expectancy.

Probably the most important factor in seeing this technique expand in its use is precedence. Other important factors in promoting its adoption are those seem to be operating effectively in similar structures and a more realistic modelling of life expectancy.

ICCP USE IN OLD STRUCTURES

ICCP is an expensive technique and also has a monitoring requirement, as galvanic systems and hybrid systems should also have. The types of anode and power supplies used to date have not really been made with a cold logic but more with what the specifier is familiar with. Thus, different countries have adopted different solutions to the same problems. Perhaps the most comprehensive selection procedures are in Germany. The German system is where an ICCP product (such as an anode or reference electrode) has to be tested by an independent university special-ised in this subject and found to be equal or demonstrably better than an existing product that has previously been tested and is being commercially used. This principle and the long-term moni-toring contract being mandated as part of the construction award has led the German installations to be almost universally success-ful. This is in direct contrast to the American experience, where cost cutting, poor specifications and poor installation practice have made failure the normal. Now that ICCP is coming back in North America, the same thing is apparently happening again, even with clients who understand how the technique should be applied. One of the reasons for this is that CP is normally only a small part of the project and is subbed out by the main contractor to the cheapest specialist subcontractor. To win this subcontract, they use the cheapest unskilled labour and the cheapest materi-als for all the system components. When the project is finished, there is very rarely any monitoring or maintenance requirement. This has meant that several ICCP systems have been abandoned shortly after commissioning and sometimes before.

There are signs that in several countries, with increasing com-petition and sizes of the ICCP projects, the standards and material specifications are being reduced in order to win projects. This is likely to have a significant reduction on the durability and effec-tiveness of the ICCP system, which are installed now and in the near future, and is quite likely to diminish its popularity in the long term.

GALVANIC ANODES IN OLD STRUCTURES

There has been a huge increase in the availability of this type of anode as the very large construction chemical companies have almost all introduced galvanic anode ranges in their repair offer-ings. The range in quality and sophistication about what is being offered is quite dramatic. It would be in the interests of these commercial construction chemical companies to define what these anodes are suitable for, that is, preventing incipient anodes, cathodic augmentation or cathodic current application (and hence corrosion rate reduction). Many of the brochures being published are showing that these anodes are compliant with standards that specifically exclude them. One of the many unknowns with these anodes is their medium- and long-term effectiveness and longevity.

One of the most urgent tasks for an orderly use of this technology is to develop a realistic test scenario where the different anodes can be tested over an extended time period with an indication of their current output, throwing power and how performance reduces with the passage of time and varies with temperature. This can then be coupled to more realistic current demand and current distribution requirements in order to affect and perhaps stop the corrosion process.

IMPROVED ANODES

There are several new anode systems coming to market. Perhaps the most significant is a new range of paint anodes which seem to have overcome the weaknesses of the old water-based paint systems which had significant current dumping issues and poor adhesion in wet conditions.

There are also improved galvanic anode systems in development, which have the potential to meet the traditional criteria for CP.

Some truly awful anodes have also been recently launched in the last few years, and surprisingly, some of these are being backed by large companies who should know better.

CHEAPER ELECTRONICS

The price of mass produced electronics is continuing to drop, and there are more specialised measuring circuits becoming available. This has led to the zone sizes shrinking and the number of sensors increasing. There is also a trend for more distributed systems. In the future, more of these systems could be self-powered, increasing their greenness credentials and reducing the wiring costs. The increased connection of systems will continue to improve in the near future meaning that almost all the systems installed will have remote operation and data transfer.

IMPROVED SENSORS

Presently, the type of sensor used to define the effectiveness of CP has shrunk to reference electrodes alone. This in some ways is a good thing because these sensors have been proven to be durable and stable, but they also have a significant weakness in that they are not actually directly telling whether the corrosion is stopped or not. When CP is applied to the concrete, there will be chemical changes, which will indicate whether passivation has been achieved. Trying to monitor these changes will provide direct information on when the corrosion process has been halted. A big problem is that in placing sensors in an old structure, the original environment may be disturbed, and innovative ways of bypassing this problem are required.

CHAPTER 6

How Cathodic Protection Works in Reinforced Concrete

INTRODUCTION

Cathodic protection (CP) of reinforced concrete structures has become increasingly popular and is now being routinely installed on many structures around the world. The most popular cathodic protection systems have their anodes placed on the surface of the concrete, in slots or in holes. Generally, the couplant to electrolytically connect the primary anode to the concrete substrate is grout, mortar or sprayed concrete with a hydraulic cement base. This bonds to the original concrete, which is in ionic connection with the steel reinforcement and provides a continuous path for ionic current flow from the anode to the reinforcement.

At the primary anode surface, there is a chemical reaction in the case of a galvanic anode or with DC power for an impressed current cathodic protection (ICCP) anode producing a surfeit of electrons. These charges are carried as ions through the cementitious materials. When the ions reach the steel reinforcement, there is a chemical change at the surface and the current is again passed by electrons. In this section, the three processes are looked at more closely and their similarity and divergence with traditional CP is discussed.

CURRENT DISCHARGE AT THE PRIMARY ANODE

This is where the anode has a surfeit of electrons, the electrical charge of which is carried through the cementitious electrolyte by an ionic flow. At this change from electronic flow to ionic flow, there will be a reduction in alkalinity. This is because of the production of hydrogen ions (oxidation of water) or chlorine gas (oxidation of chloride ions) at the anode surface.

$$2H_2O = O_2 + 4H^+ + 4e^-$$

$2Cl^- = Cl_2 + 2e^-$ going on to form hypochlorous acid

For traditional CP systems protecting steel in water, it is considered that one of these two reactions occurs at the anode, depending on whether the water surrounding the primary anode is saline or not. If there is chloride, the chlorine reaction will be favoured. In freshwater, the oxygen production reaction is favoured. In concrete, the situation is likely to be more complex, as there are many compounds, which may react, and this has yet to be fully researched but some relevant testing to date is presented. The primary anode also has an influence, for example, certain components of mixed metal oxide (MMO) anodes favour different reactions. For example, RuO_2 favours oxygen production, while IrO_2 favours chlorine production. The ratio of the oxygen to chlorine reaction possibly depends on the relative concentrations of water to chloride and the potential difference at the anode to grout interface. Another factor controlling the generation of chlorine or oxygen is the magnitude of the current density applied at the anode surface. The MMO ribbon commonly used for CP in concrete is an IrO_2/TaO_2 mix, which favours chlorine production (Shrier, 1998).

ACIDIFICATION AT THE ANODE

When current is passed at the anode, acid is generated by water oxidation and/or chlorine gas dissolution, and this could cause damage to the cementitious couplant. Because of this, there are, in several standards, for example, EN ISO 12696 (2012), current density limitations placed on the anode. These limitations have profound implications on the economics of CP, as it could increase the number of anodes required by a factor of four or more.

CP of reinforced concrete was first tried by the Americans in the late fifties (Stratfull, 1959), and it had expanded to preventing further corrosion of reinforced concrete bridge decks in the late 1960s and early 1970s. They found that the high resistance and limited cover depths of the concrete placed significant limits on the output of an anode and its current spread. They also found that over a certain current density, damage may occur (Bennett et al., 1993) to the concrete or couplant mortar adjacent to the primary anode. Researchers first noticed this problem in site observations of impressed current slotted systems on bridge decks and undertook limited laboratory experimentation with a MMO ribbon anode cast into the concrete. This research (Bennett et al., 1993) provided the data that were used as the basis for the anode current density limitations in the present European standard (EN ISO 12696, 2012). This experiment used MMO ribbons cut into slots in the concrete surface, an arrangement that is known for its uneven dispersion of current. Acidification was deemed to have occurred if there were any signs of cementitious softening or discolouration along the ribbons' length. This effect was found to occur at an applied current density higher than 110 mA/m^2 of the anode surface area.

As discussed in Chapter 4, acidification at the anode/cementitious couplant interface is a serious durability problem for certain anode systems, as a feedback loop can develop with the eventual failure of the anode at this location. This is particularly important in aggressive environments such as marine installations, and the anode system should be designed to prevent this occurring.

SUMMARY OF DANISH, NORWEGIAN AND AMERICAN ACIDIFICATION TRIALS

1. Three different cementitious grouts (designated A, B and C) were tested at CP International in Denmark (Gronvold, 1996) at extremely high current densities for 6 months indoors. The test rig had two commercially available MMO discrete anodes embedded in a test mortar tube made from the cast grout. This tube of 12 mm diameter was placed in a salt water bath with a steel cathode. This test was at an anode current density of 1,600 mA/m^2. At the conclusion of the test, specimen group A had minor damage, specimen group B had some damage and specimen group C was destroyed.

2. A purpose built outdoor test facility was constructed by a discrete anode manufacturer using an European Union research subsidy to allow detailed measurement of impressed current anodes' performance and current spread. This had a water storage tank cast into the top of the structure, so high anode currents (700 mA/m^2) could be passed at normal voltages (4–5 V). These trials were run for up to 8 years with core examinations showing limited or no acid damage at the interface between the special grout (this had fine particles to give good flow characteristics and ionic conduction enhancers, and was based on the Group A grout above) and the MMO-coated titanium tube (Gronvold, 1998).

3. An early test (Danish Road Authority, 1994) of acidification at extremely high anode current densities was made in the middle 1980s. Here, the couplant was a graphite paste,

which is electronically conductive, directly linked with the parent concrete. The anode test was at 4,000 mA/m² at a DC voltage of between 60 and 70 V for 1 year. The anode-to-concrete interface was examined by the Laboratory of the Danish Road Authority. They found that about 2–3 mm of the concrete abutting the graphite was acidified but no concrete material was missing and also no graphite had been oxidised. The researcher found on this and other tests that the critical factor was to have no voids against the primary anode surface which can fill with water.

4. A trial (Norwegian Highways Authority, 2003) made an evaluation of acid damage to the anode interface. Its conclusion was that the mortar-surrounding MMO anodes with special grout suffered little or no acid damage over an operating period of 8 years at 450 mA/m². This finding has been replicated on literally thousands of ICCP systems around the world, some of which have been operating for more than 25 years at this or slightly lower anode current densities.

5. Testing of Elgard discrete anodes by J. E. Bennett Consultants (the same person who did the original experiments) in Ohio, the United States, for Eltech's range of MMO discrete anodes, began in 1996. Anode current densities of between 387 and 775 mA/m² were used, and the anodes and surrounding mortar autopsied after 235, 404 or 570 days under current. The removed core was split, and the anode and surrounding grout were examined for damage. The damage being looked for was anode reaction product (ARP), which is the softening and discolouration of the cement paste. It should be noted that this ARP does not mean that the anode had stopped functioning but merely that the appearance and structure of the grout had changed. Table 6.1 was compiled using the data from two reports (Bennett et al., 2007, 2008) detailing this research programme.

TABLE 6.1 Research by Bennett on Acidification with Impressed Current Anodes

Anode No.	LiNO$_3$ Added to Mortar	Current (mA/m^2)	Time (Days)	ARP Outside	ARP Inside	Overall Condition
1	N	387	235	None	None	Excellent
6	Y	387	235	20%	None	Good
7	N	387	235	None	None	Excellent
10	Y	516	235	Extensive	Extensive	Poor
3	N	387	404	Less than 20%	Significant	Excellent
11	Y	387	404	None	None	Excellent
9	Y	516	404	None	Significant	Excellent
5	Y	775	404	20%	Significant	Good
2	N	387	570	30%	Significant	Good
4	Y	387	570	None	None	Excellent
12	Y	387	570	Less than 20%	None	Excellent
8	N	516	570	None	None	Excellent
13	Y	775	570	Extensive	Extensive	Fair

From Table 6.1, it can be read that at an anode current density of 387 mA/m^2, all the anodes inspected showed minimal damage to the cementitious grout. At 516 mA/m^2, some mortars showed significant damage and some did not. At 775 mA/m^2, there was some damage to one mortar and little to another. There was no correlation between the time when the current had been passed and the amount of damage. There were very significant differences in performance between mortars at the same anode current density. It should be noted that ARP was noted on the inside surface of the anode in certain autopsied anodes (the anode was perforated and hollow, so mortar was on the inside and outside of the anode). This implies that a significant proportion of the current was provided from the inside surface of the anode. The grout used was Sika 223. Some of these samples had lithium nitrate added to the grout to improve the acid resistance, and this is documented in the table with a Y for yes and N for no. In the overall conclusions, it was stated that there was no significant difference in ARP formation between standard grout and the grout containing added lithium nitrate.

SUMMARY

Experiments to test acidification at between 3 and 36 times higher current densities than that which is permitted by present standards have been undertaken by various researchers in different countries. They have found that there is some correlation between higher current density and more acidification, but even in regulated laboratory experiments, there was a great inconsistency to the results, which implies that another factor is more important in causing acid damage than the current density. As the samples' chemistries were constant, the only variable possible is that the mixing and subsequent voidage (holes in the set grout matrix) in the samples was unequal. This voidage had a dramatic effect on the performance of the specimen.

CALCULATION OF ACID FORMATION AT THE PRIMARY ANODE

If one assumes that the anodic reaction occurring at the encapsulating cementitious material is the formation of hydrogen ions, then a life expectancy with a commercial cementitiously enveloped discrete anode system at its recommended maximum output can be calculated for a 12-mm-diameter hole of 100 mm length;

Mass of the cementitious material in hole = 22.5 g.

The amount of alkali in the grout is assumed to be 19%.

Molar weight of $Ca(OH)_2$ = 74 g, so the 100-mm-length hole = 22.5/74 × 0.19 = 0.058 mol.

Operating at 450 mA/m² computes to an output of 2.39 mA for a 100-mm length.

The number of amp hours available in the alkali = 0.058 (mol) × 26.8 (A h/mol) = 1.55 A h

Assuming a 50% electrochemical efficiency then at this output, all the cement paste will be consumed in = (1.55 × 1,000/2.39) × 1/0.5 = 1,292 hours = 54 days.

COMPARISON BETWEEN PREDICTED AND ACTUAL ANODE MORTAR DEGRADATION

The acidification predicted by the above-mentioned electrochemical calculations did not occur in any of the practical trials of the mortars outlined above. Furthermore, extremely high levels of current output were applied to some of the trial samples, and no, or little, discernible damage was noted after 8 years of testing. This implies that the calculation for anode cement degradation is an over simplification of a complex process. This finding was confirmed by additional research by Bertolini et al. (2013) who used colour-pH indicators to show that acidification around the anode had occurred. The amount of acidification they had expected in accordance with Faraday's law did not occur. The actual amount of acid they found was about 10% of the level predicted. It was suggested that the hydroxyl ions produced at the cathode migrated to the anode and neutralised the acid. Further research work is described by this source where thin sections of the anode/mortar interface were taken and studied using an optical microscope and a scanning electron microscope. In this study, it was suggested that mechanisms other than hydroxyl migration mitigated the predicted level of acid damage. Further tests were done on ICCP system anodes which had passed current at 20 mA/m^2 for up to 6 years. Again there was less acid formation than that calculated, and it was stated that there were other processes at work and these are insufficiently understood. In conclusion, it stated that 'acid formation at the anode, at least at moderate current densities, is not critical for the service life of CP systems. However, it may be relevant for (unintended) local high current densities'.

It appears significant that in different trials described above (in 1 and 5) at identical current densities, we are getting diverging outcomes (from little damage to sample destruction) for no obvious reason. During the trials, it was noted that in certain circumstances, anodes were producing chlorine gas on a continuous basis, gas that could be smelled, seen (a greenish yellow gas) and analysed, at other times hypochlorous acid was being produced. This took place at the low voltages (say 3 to 5 volts) and currents used in the commercial CP of reinforced concrete. The discharge of chlorine means that the oxygen-producing reaction of $2H_2O = O_2 + 4H^+ +$ electrons$^-$ has not occurred and the alternative of $2Cl^- = Cl_2 +$ electrons$^-$ has. Of course, both reactions can occur simultaneously. It also means that the chloride has moved from the concrete through the mortar (which is chloride free) and arrived at the primary anode surface. This chlorine- and hypochlorous-producing reaction gives much less acidification than the water oxidation reaction (Shrier, 1998). For this chlorine-producing reaction to occur, there must be plentiful resupply of chloride ions from the surrounding matrix. This can only arise if the surrounding concrete is chloride contaminated, and this may account for the large difference in behaviour of certain specimens under the same experimental conditions, that is, if the oxygen production dominates, there will be more acidity produced with consequential cement paste damage, whereas if chlorine gas is produced, there will be less damage.

Another factor that appears to be critical to the performance of the mortar is voidage where there is incomplete encapsulation of the anode. This voidage could mean that ions flow by a different mechanism through the void filled with pore water as through the cement matrix (see Chapter 1). It is known by a discrete anode manufacturer that poor grouting (i.e., voidage) has a huge effect on the durability of the anode system.

IONIC MOVEMENT THROUGH WATER

In traditional CP, in sea water, for example, the ions flow through the solution at a rate which is determined by the size of the potential difference with only a slight directional movement imposed on their random motions. The cations are attracted towards the cathode with the anions moving towards the anode. The ionic mobility of these cations and anions is measured practically with a potential difference of 1 V and a distance of a metre. Water making up the solution is not expected to move significantly, that is, electrical endosmosis.

It is known from practical experimentation that most ions (chlorides, sulphates, calcium, sodium) travel at roughly the same speed whatever their size with two exceptions (Barrow, 1973). These are H^+, which is six times faster than a typical ion and OH^-, which is three times faster. This is probably due to a proton transfer mechanism. In aqueous solutions at a potential of 1 V/m, it will take a typical ion 30 minutes to travel 1 metre. There are several hypotheses such as Stokes radius, degree of solvation and some newer ones, such as solvation sheath, which are used to try and explain why ionic mobility does not directly follow the size of the ion. Most of these are simple approximations that are ascribing macroscopic properties to a molecular scale.

We would expect sodium, calcium, magnesium and potassium to all be attracted towards steel which is being cathodically protected. This is seen to occur on CP systems applied to protect steel in sea water, which forms a whitish film on the surface of the exposed metal. Analysis has shown this to be a film of predominantly magnesium hydroxide. This salt is deposited because of its lower solubility compared to the other metal hydroxides.

IONIC MOVEMENT THROUGH CONCRETE

When concrete sets, the hydrated cement consists of solid hydration products plus the water, which is held physically or is adsorbed on the large surface area of the cement hydrates. This is called gel water and is located between the solid products of hydration in gel pores.

The solid products of hydration occupy a smaller volume than the original solid and thus there is residual space in the structure, which are called capillary pores. These can be empty or full of water and are about 1,000 nm in width (Neville, 1995).

The capillary pores are big enough to allow the double layer conduction described in Chapter 1 to occur but the tortuosity of the pores transit paths will account for a considerably slower transport for the ions.

When an electrical field is applied, as in cathodic protection, the water molecules in the capillary pores will align themselves according to their polarity. This is because the water molecule is not electrically neutral and is thus polar. There is a small net negative charge near the oxygen atom and partial positive charges near the hydrogen atoms. This causes an inner layer of bound water to attach to the pore wall and then an outer layer, which is free to move. This is the so-called electrical double layer. The immobile surface charge surrounding the concrete particle in turn attracts a cloud of free ions of the opposite sign creating a thin (typically 1–10 nm) Debye layer next to it. The thickness of this electric double layer is determined by a balance between the intensity of the thermal (Brownian) fluctuations and the strength of the electrostatic attraction to the substrate. In the presence of an external electrical field, the fluid in this charged double layer acquires a momentum, which is then transmitted to adjacent layers through the effect of viscosity. In this outer layer, the water can flow taking dissolved salts along with it.

CONCENTRATION PROFILES OF IONS BEFORE AND AFTER CATHODIC PROTECTION

Experimental work (Sergi & Page, 1995) looked at how the profiles of four ions varied after a CP circuit was made for a month at an average current density of 350 mA/m² and a potential of the cathode of −850 mV with respect to a silver calomel electrode. A graph of the data is shown in Figure 6.1.

This shows that there is significant movement of certain ions over the time period compared to the control specimen where no electrical field had been applied. This is on the right-hand side of the figure. The movement of ions is most pronounced in the 10 mm closest to the anode and the 10 mm closest to the cathode. The largest movement is of hydroxyl ions with the halving of its concentration at the anode and increase by approximately 50% at the cathode.

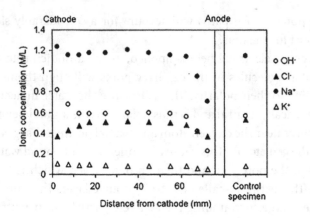

FIGURE 6.1 Profiles of ionic concentration measured between an anode and a cathode.

The movement rate of the sodium ions almost matches that of the hydroxyl ions can be explained by the primary source of moment being electro-endosmosis with water moving to the cathode and away from the anode.

The anions (K^+ and Na^+) behave very differently, with sodium being removed from the anode and congregating at the cathode. The potassium ions distribution has not been affected by the imposition of the electrical field. Both these cations' chloride salts have virtually identical solubility in water, so this is not an explanation for their different behaviour which according to present chemical principles should be similar.

The findings of this experiment are different to that which would be expected from the behaviour of ions in an aqueous solution, but what can be seen is that there is a significant movement of certain ions in a relatively short space of time (1 month). Perhaps the most critical is the movement of chloride, as this will reduce the resistance at the anode both directly and by the fact that the salts are hygroscopic. Secondly, the presence of chloride will change the anodic reaction from water oxidation to chlorine production, which is significantly less acid producing.

CHANGES TO THE STEEL AT THE CATHODE

With the transport of hydroxyl to the steel reinforcement it is likely that there will be a significant change to the surface conditions of the steel. Some research has been done on this, but it could be argued that this change is likely to be different between micro and macro corrosion, that is, general corrosion and pitting. Apart from the increase in alkalinity, the other large change is likely to be the diminution of chloride by it flowing in the electrical field towards the anode (and repelled from the cathode), though this is significantly influenced by the anode geometry, the steel reinforcement layout and the location of the chloride contamination.

In the early days of development of CP, there was significant concern that the increase in alkalinity, which would be generated by cathodic protection could trigger alkali silica reaction and/or reduce the bond strength. This is the mechanical connection between the steel and concrete matrix. Limited evidence was found that either of these two factors was significant at the current and voltage levels being applied, but it did introduce the notion that the biggest single effect of cathodic protection was to increase the pH of the steel reinforcement generally across its surface and also in a corrosion pit.

This was investigated (Pacheco et al., 2012) by an extended study where some samples had cast in chloride, some had salt applied on the outside and some were cycled in a carbon dioxide-rich environment. The samples were then left outside for 10 years. There was clear evidence in some of the samples that active corrosion was occurring at this time. The samples were then cathodically protected and finally destructively examined. From these measurements, there was evidence that some parts of the specimens were corroding at a greater rate than others, which was a clue about orientation of the steel being important. In this trial, they estimated the cathodic protection at 1.4 V with an initial current density of 30 mA/m^2 dropping to 10 mA/m^2 of steel reinforcement area within 24 hours. The experiment concluded that the acidity in the pits was neutralised within 8 hours, and this allowed the re-passivation of steel. The current densities in these experiments preclude the use of galvanic or hybrid systems.

One of the most important mechanisms explaining how cathodic protection works is that as pitting or general corrosion is stopped at more corrosive locations, the resistance of this ionic pathway increases causing the current to flow to another locations. This effect is commonly seen when a cathodic protection system is energised in a relatively dry atmosphere. When in a marine environment, often the amount of current required stays relatively stable over an extended period of years.

CONCLUSIONS

Perhaps the most interesting insight that the data presented in this chapter is the notion that concrete is a closed environment where movement of ions through a dense and impenetrable structure is extremely slow with significant movement over decades is a gross simplification of a much more complex and dynamic mechanism. Under the small potentials (typically 5 V) used for ICCP, there is significant movement of ions within hours of this potential being applied. The movement of ions appears to be much more specific than would be predicted from aqueous testing for reasons which are yet to be explained. What has been demonstrated that both the hydroxyl and chloride ions are very mobile and that both these ions' movements are likely to have a profound effect on the application of CP to a reinforced concrete structure. The chloride will be moved away from the steel reinforcement, thus preventing further breakdown of the passive oxide film, also the chlorides will be moved to the anode, which is beneficial for a number of reasons. The chloride reduces the electrical resistance of the mortar surrounding the anode both by its presence and its attachment to water molecules. The chloride also promotes a lower acidity anode reaction. The hydroxide movement to the steel reinforcement has two benefits in that it increases the pH of the steel surface and promotes film formation, which will both act as a barrier to chlorides and electrostatically move the chlorides away from the steel surface.

REFERENCES

Barrow G, *Physical Chemistry*, (1973), McGraw Hill, New York 3rd edn.

Bennett J et al., *Cathodic Protection of Concrete Bridges: A Manual of Practice*, (1993), Transportation Research Board, Washington DC, USA, SHRP-S-372, p. 28.

Bennett J et al., Testing of Stargard discrete anodes for concrete, NACE 2007 paper. NACE International, Houston, USA.

Bennett J et al., Eltech discrete anode tests 570 day autopsy and operational report, June 20, 2008. Document issued to industrial client.

Bertolini L. et al., *Corrosion of Steel in Concrete*, (2013), Wiley-VCH, Weinheim, Germany. 372, ISBN 9783527331468.

Dansk Road Authority report, (1994). Document issued to appropriate agencies in Denmark.

EN ISO 12696, *Cathodic Protection of Steel in Concrete*, (2012), British Standards Institute, London, UK.

Gronvold F, Gronvold & Karnov internal trial document, (1996).

Gronvold F, Gronvold & Karnov internal trial document, (1998).

Norwegian Highways Authority (Statensvegvesen), (2003). Translation-use of CP on highway structures Document issued as technology note to Norwegian relevant authorities.

Pacheco J et al., Short term benefits of cathodic protection of steel in concrete, in *Concrete Solutions*, (2012), Taylor and Francis, London, UK, p. 147, ISBN 9780415616225.

Sergi G and Page L, Advances in electrochemical rehabilitation techniques for concrete, *UK Corrosion*, (November 1995), 95, 21–23.

Shrier L et al., *Corrosion*, (1998), Butterworth Heinemann, Oxford, UK, vol. 10, p. 66–67, ISBN 0750610778.

Stratfull R, *Corrosion*, (1959), NACE International, Houston, USA, 15(6), 65–68.

CHAPTER 7

Defects with Existing Standards

PERFORMANCE CRITERIA HISTORY

Cathodic protection (CP) has been successfully used for more than a 150 years on ships, where the high conductivity of sea water made it easy to pass the current. CP came ashore over a 100 years ago to protect steel in the ground, and this was found to be achievable despite the lower conductance of the soil electrolyte compared to sea water. Robert Kuhn has become known as the 'Father of Cathodic Protection' in America, and he installed the first impressed current installation in 1928 (Baeckmann et al., 1997) on a long-distance gas pipeline in New Orleans. By experimentation, Kuhn found that a protective potential of −850 mV with respect to a copper/copper sulphate electrode provided sufficient protection against any form of corrosion. He found that this level of protection was being obtained at a current density of 10–20 mA/m² for the steel of uncoated pipe. An early way of measuring effectiveness of CP was the incidence of pipe bursts. This was because electrochemical corrosion was responsible for the pipe failures in aggressive soils. The incidence of these failures could be easily recorded. After CP was applied (Peabody, 2001), the number of pipe bursts was shown to reduce to zero incidences.

The direct negative criteria was clear and easy to practically measure, and over 60 years later (BS 7361, 1991), it had been modified only slightly from −850 to −950 mV with respect to a copper/copper sulfate electrode in anaerobic environments where bacteria could increase corrosion rates significantly. There was also mention of eliminating the IR drop error by switching off the power supply or disconnecting the anodes to remove the potential drop component produced by the flow of current through the electrolyte. The IR drop is the voltage dimunition between the two ends of a conduction phase during current flow. The voltage drop across any resistance is the product of current (I) and resistance (R). By switching off the current, this voltage drop is removed. By this time, direct evaluation of the corrosion process in saline water had been commonly undertaken directly demonstrating that the corrosion process was stopped on steel with CP applied to the levels in the standard. This was backed by weight loss measurements of the steel, which showed that no metal was being consumed.

By 1991, CP was being directly applied to atmospherically exposed reinforced concrete structures and BS 7361 had a section and an appendix on criteria for steel in concrete, and this suggested a depolarisation criteria of 100 mV after up to 4 hours could also be appropriate to demonstrate that corrosion had been stopped by CP. It was suggested that this was measured with a portable reference electrode on the surface of the concrete at many locations. The reason that this new criteria was introduced was that it was found difficult to reach the −850-mV criteria on many occasions in both laboratory experiments and on site on exposed reinforced concrete specimens and structures. It was reported in this standard that successful CP appeared to have been achieved at a current density of between 5 and 20 mA/m^2 of the steel reinforcement surface area. This 100-mV criteria was tested for validity in the United Kingdom by the Transport Research Laboratory (TRL) with a final report (McKenzie, 1993) describing several years of work on multiple research programmes. In one of the large-scale experiments, they made multiple specimens, allowed the steel reinforcement to corrode, and then tried to stop further CP. Here, due to anode problems, the amount of current applied varied as did the depolarisation levels. It was found by destructive evaluation that where the depolarisation level achieved was more than 100 mV, little further corrosion occurred, although this was not universally the case. This and other series of experiments by TRL were instrumental in validating the criteria for concrete given in BS 7361.

In the BS 7361 standard, no mention of using galvanic anodes directly in reinforced concrete was made, presumably because they were thought to be unsuitable for this application.

PERFORMANCE CRITERIA PRESENTLY

The present situation in Europe is that we have a recent standard (EN ISO 12696, 2012) which is specifically for the 'Cathodic Protection of Steel in Concrete'. This is a much larger document than the previous British Standard (BS 7361, 1991) and runs to 45 pages. This standard lays out in more detail what is required for the design, components of a CP system and monitoring criteria. One of the many things it says is that 'galvanic anode systems may be used without monitoring systems or methods to measure their performance. Such systems do not comply with this International standard'. The majority of galvanic anode systems are installed without any thought of monitoring their performance (the anode tail insert is directly attached to the steel reinforcement beside it) through current output measurement, so this means that presently they are specifically excluded from this EN ISO standard as shown in Figure 7.1.

One of the big problems that this present standard faces is that it is attempting to codify good practice and acceptable performance on a technique that is commercially being used more widely than ever before but is not fully explained or understood.

FIGURE 7.1 Galvanic anode electrically connected directly to reinforcement. (Courtesy of Mapei).

The document tries to tread a difficult path of being inclusive to the many commercial interests while also trying to establish the best practice. This is illustrated most clearly in the criteria of protection as this defines the success or otherwise of a reinforced concrete CP installation and so is obviously of great commercial importance to installers and material manufacturers. First, there is a development of the criteria from BS 7361 (1991) so that it is more flexible with the 4-hour depolarisation period extended to a maximum of 24 hours, and the new 150-mV criteria introduced over a time period of 72 hours are now acceptable as evidence that CP was being achieved. These criteria, it was stated, were not necessarily supported by theoretical considerations but are a non-exhaustive, practical series of criteria to indicate polarisation which will lead to the maintenance of or re-establishment protective conditions for the steel within concrete.

To obtain any of these criteria with galvanic anodes could be very difficult, particularly if they have only a small surface area (see Chapter 4) as their individual current outputs will be so much less than an impressed current system, so an additional non-normative note was added specifically for galvanic systems whose current can be measured. In this note, it was written that 'the steel corrosion rate may be estimated by inserting the applied current density and steel potential shift into the Butler–Volmer (BV) equation. The applied current density may be obtained from the current delivered from a small segment of the anode system and an estimate of the steel potential decay measured at the same anode segment'. Passive steel, it was written, is indicated by a corrosion rate of less than 2 mA/m^2 and preferably less than 1 mA/m^2. Why galvanic anodes should have different criteria to ICCP does not appear to be logical.

This inclusion has been used on many occasions to indicate that satisfactory CP has been obtained with galvanic anodes, but it is a step change from the previous practical series of criteria mentioned above, so what is it and is it valid?

The BV equation is a mathematical construction from the 1930s that describes how the electrical current passed from an electrode depends on the electrode potential. It was derived as a combination of the Nernst equation devised in 1900, which provides a reverse potential as a function of the ionic concentration in solution, and the Tafel equation. This Tafel equation was developed through experimentation in an aqueous media in 1905, and the formula derived is from the fact that at low currents, the applied potential is proportional to the log of the resulting corrosion current.

The BV equation shown below is a large equation which links several of the parameters such as faradaic current, electrode potential, concentration of the reactant, concentration of the product and temperature. Another way of describing this equation is that it is a standard model that can be used to describe the current–overpotential relationship of an idealised electrode at a specified temperature, pressure and concentration of a single aqueous reaction.

$$I = A \cdot j_0 \cdot \left\{ \exp\left[\frac{\alpha_a nF}{RT}\left(E - E_{eq}\right) \right] - \exp\left[-\frac{\alpha_c nF}{RT}\left(E - E_{eq}\right) \right] \right\}$$

This can be given in a more compact form:

$$j = j_0 \cdot \left\{ \exp\left[\frac{\alpha_a nF\eta}{RT} \right] - \exp\left[-\frac{\alpha_c nF\eta}{RT} \right] \right\}$$

where
- I: electrode current, A
- A: electrode active surface area, m^2
- j: electrode current density, A/m^2 (defined as $i = I/A$)
- j_0: exchange current density, A/m^2
- E: electrode potential, V
- E_{eq}: equilibrium potential, V
- T: absolute temperature, K
- n: number of electrons involved in the electrode reaction
- F: Faraday constant
- R: universal gas constant
- α_c: cathodic charge transfer coefficient, dimensionless
- α_a: anodic charge transfer coefficient, dimensionless
- η: activation overpotential (defined as $\eta = \left(E - E_{eq}\right)$).

To explain what is being stated above more directly, more than 100 years ago, experiments were undertaken which put a potential sensor (reference electrode) close to a piece of metal in an aqueous solution and found that when you passed current through this metal, its potential changed. The amount the potential changed was related to the amount of current passed and also to the passivity of the metal. It was found in practical experimentation on some metals that for a small region (typically a range up to ±20 mV) around the natural potential (with no current applied) when the current is logged against the voltage, it would follow a roughly straight line. This line is called the Tafel slope. As the voltage is increased, this line tends to deviate further from a straight line as shown in Figure 7.2.

This can be compared to a small CP system with a reference electrode close to the steel. When you pass current from the anode, the potential of the steel measured by the reference

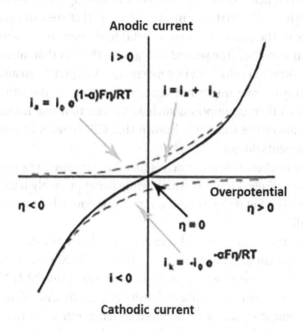

FIGURE 7.2 BV representation of electrode current versus potential. The black line at zero current is at the Nernst point. The straight Tafel slopes are at the upper and lower parts of the black lines from the origin. The dotted lines show the overpotentials.

electrode changes. The amount it changes is proportional to the current passed and the passivity of the metal. In essence, the BV note in the standard (EN ISO 12696, 2012) is attempting to use this to get a measure of the passivity of the steel reinforcement. This is also the basis on which linear polarisation resistance (LPR) measurement equipment for reinforced concrete works. In this, a current is passed through the concrete and the difference in potential measured on the steel reinforcement using a portable reference electrode. This LPR testing has only been moderately successful as the situation in concrete is complex and variable, so the test data obtained by this method is commonly verified by a limited destructive examination of the reinforcement.

This BV criteria is unlike the previously used assessment methods and is much more complex. Also it is making some significant assumptions. The first and most dubious is that steel in concrete behaves in the same way that a metallic element in an aqueous environment does. The second is it ignores the fact that micro and macro corrosion cells exist in concrete (see Chapter 1), so an average corrosion rate might not adequately represent the situation. There are other assumptions such as the reaction rate increasing with temperature when it is known that CP currents in aqueous environments do not.

What in effect is being stated is that if the corrosion rate is very low, you will need less CP, which while being probably true, begs the question that if the corrosion rate is very low, why do you need CP at all?

The output criteria for achieving CP has been broadened from the old British Standard (BS 7361, 1991) of 5–20 mA/m^2 of the steel reinforcement surface area. The current standard (EN ISO 12696, 2012) requires a current density of 2–20 mA/m^2 of the steel reinforcement surface area required in temperate conditions. No guidance on what current density might be required in tropical, subtropical or arctic conditions is given.

A new category of cathodic prevention has been established (EN ISO 12696, 2012) with a current density of 0.2–2 mA/m^2 of the steel reinforcement surface area. This cathodic prevention category has been introduced, as it is believed that where corrosion is prevented from initiating rather than being stopped, there is a lower current density requirement. This does seem to be the case for CP of steel in soil and sea water. Whether this is the case for steel in concrete is yet to be determined because of the uneven nature of the corrosion initiation and propagation which has been found in steel in reinforced concrete (Angst et al., 2011).

What is striking in both of these CP current requirements presently being used is the variability being shown here with both these limits having a factor of difference of one decade or ten times. This suggests great uncertainty on what constitutes CP or cathodic prevention and an ignorance of the corrosion process that they are attempting to stop.

PERFORMANCE DATA OF VARIOUS CP SYSTEMS

As an illustration of this is the operating data from three types of galvanic anode systems and an impressed current system showing the disparity that exists between the amount of protection being afforded. All the potentials given are relative to a manganese oxide reference electrode.

- **Project description**: precast concrete beams inside a cooling tower

  - Anode type: discrete anode – impressed current

  - Anode spacing: 400 mm centres (6.25 anodes per m^2 of concrete)

  - Output: constant voltage 4 V

  - Current density after 1 month: 22 mA per m^2 of steel reinforcement

  - Natural potential: average of steel reinforcement measured by eight reference electrodes −540 mV

- Instant off: average potential after 1 month −830 mV
- Average 24-hour depolarisation: 210 mV
- Criteria: pass with negative potential and depolarisation level
- Comments: concrete saturated but still significant depolarisation.

- **Project description**: cast in situ support beam of motorway
  - Anode type: sheet zinc with gel adhesive – galvanic
  - Anode spacing: over all external surface
  - Output: voltage 0.5 V
  - Current density after 1 summer month: 1.2 mA per m^2 of steel reinforcement
  - Natural potential: average of steel reinforcement measured by four reference electrodes −380 mV
  - Instant off: average potential after 1 month −410 mV
  - Average 24-hour depolarisation: 36 mV
  - Criteria: fail with negative potential and depolarisation level pass with BV equation
  - Comments: concrete outdoors but weather protected by deck so dry.

- **Project description**: repair of deck in a multi-storey car park
  - Anode type: zinc rods in cementitious repair – galvanic
  - Anode spacing: 400 mm centres (6 anodes per m^2)
  - Output: constant voltage 0.5 V
  - Current density after 1 month in autumn: less than 0.2 mA per m^2 of steel reinforcement

- Natural potential: average of steel reinforcement measured by two reference electrodes −410 mV

- Instant off: average potential after 1 month −405 mV

- Average 24-hour depolarisation: less than 2 mV

- Criteria: fail with negative potential and depolarisation level possible pass with BV equation

- Comments: concrete dry.

- **Project description**: Abutment wall splashed by cars

  - Anode type: zinc mesh in alumo-silicate compound – galvanic

  - Anode area: about 70% of concrete surface anode area

  - Output: constant voltage 0.5 V

  - Current density after 1 year in June: 5 mA per m² of steel reinforcement

  - Natural potential: not given

  - Instant off: average potential after 1 month −457 mV

  - Average 24-hour depolarisation: less than 169 mV

  - Criteria: fail with negative potential, pass with depolarisation level

  - Comments: concrete wet and salt contaminated.

The amount of current and depolarisation being recorded in these four examples demonstrate the large difference in protection levels, which is being afforded to these different structures. There is a difference of more than 100-fold (two decades) between the highest and lowest current levels being passed by the anode system in the different examples. The zinc sheet galvanic anode has the highest possible surface area for a galvanic anode but is delivering

18 times less current than an impressed current anode; admittedly, it is in a much higher resistance environment.

SUMMARY

The latest standard (EN ISO 12696, 2012) has been stretched too far and should be retitled 'Impressed current CP of steel in concrete', and all references to galvanic and hybrid treatment systems should be removed. The criteria particularly current output should be addressed. Currents spread from individual anodes should be shown more attention.

A separate document to EN 12696 should be produced, possibly titled 'Galvanic Cathodic Augmentation of Steel in Concrete'. This second document may be difficult to produce as the amount of large companies with limited electrochemical expertise who are involved with galvanic anodes together with hybrid treatment companies is causing boundaries to be crossed.

REFERENCES

Angst U, B Elsener, C Larsen, O Vennesland, Chloride induced reinforcement corrosion: rate limiting step of early pitting corrosion, *Electrochemica Acta*, Elsevier, Netherlands, 56, (2011), 5877–5889.

Baeckmann W, W Schwenk, W Prinz, *Handbook of Cathodic Protection*, (1997), Gulf Publishing Company, Houston, USA, 3rd edn, ISBN 0884150569.

British Standard BS 7361: Part 1: *Cathodic Protection: Code of Practice for Land and Marine Applications*, (1991), British Standards Institute, London, UK.

EN ISO 12696, *Cathodic Protection of Steel in Concrete*, (2012), British Standards Institute, London, UK.

McKenzie M, Cathodic protection of reinforced concrete: effectiveness on corroded steel, (1993), project report PR/BR/41/93, Transport Research Laboratory, Bracknell, UK.

Peabody A, *Control of Pipeline Corrosion*, (2001), NACE International, Houston, USA, 2nd edn, p. 39, ISBN 1 575900920.

Index

Printed in the United States
by Baker & Taylor Publisher Services

Printed in the United States
by Baker & Taylor Publisher Services